HANDBOOK OF CRYSTAL and MINERAL COLLECTING

By William B. Sanborn

Price $2.00

Copyright ©1966 by Gembooks

Published By
GEMBOOKS
MENTONE, CALIFORNIA

Table Of Contents

Chapter 1 Becoming Familiar With
            Minerals and Crystals . . 4
Brief history of the hobby 4
Why collect minerals? 4
Publications 6
Mineralogical or gem societies . . 7
Special courses of study 8
Museums and famous exhibits . . 8

Chapter 2 Physical Properties of
            Minerals and Crystals . 14
The difference between a rock
  and a mineral 14
Physical properties of minerals . . 15
  Crystals 16
  Color 16
  Streak 17
  Photosensitivity and fading . . 17
  Luster 17
  Mineral optics 18
  Refractive index 18
  Double-refraction 18
  Asterism 18
  Hardness 19
  Specific gravity 19
  Cleavage and parting 19
  Fracture 20
  Tenacity 20
  Structural forms 20
  Electricity 26
  Magnetism 26
  Fluorescence and
    and phosphorescence . . . 26
  Luminescence 27
  Radioactivity 28

Chapter 3 Crystals, Inclusions
            and Pseudomorphs . . 29
Crystal terminology 29
The six crystal systems
  and symmetry 30
Twins 31
Parallel grouping 31
Rosettes 32
Cavernous crystals 32
Distorted forms 32
Scepter crystals 33
Inclusions; solid and liquid . . . 33
Phantoms 34
Pseudomorphs 35
  Encrustation pseudomorphs . . 36
  Alteration pseudomorphs . . . 36
  Replacement or substitution
    pseudomorphs 36

Paramorphs 37
"Skulls" 37
Fossils 37

Chapter 4 Where to Collect
            Specimens 38
Crystal environment 38
Collecting sites or sources . . . 39
  Mines 40
  Mine dumps—various types . . 40
  Tramways and loading
    platforms 42
  Open-pit mines 43
  Placer mines and dredge tailings . 44
  Quarries 44
  Outcrops 46
  Dikes and veins 47
  Cliffs 47
  Float 48
  Well cores 48
  Silicified logs 48
  Cuts and tunnels 49
Safety in the field 50
Permission to collect 50

Chapter 5 How to Collect and
            Prepare Specimens . . 52
Tools and equipment
  for collecting 52
Packing specimens in the field . . 54
Field notes or labels 55
Cleaning specimens 55
Trimming or dressing specimens . 56
Duplicate specimens 56
Specimen stability and protection . 56

Chapter 6 Specimen Size and
            Collection Types . . . 58
The importance of specimen size . 58
Size categories 58
  Micromounts 59
  Thumbnails 59
  Miniatures 61
  Cabinet specimens 61
  Museum specimens 61
Importance of localities 62
Types of collections 62
  General 62
  The Dana collection 62
  One mineral 62
  One mineral family 63
  One general chemical
    classification or
    element group 63

One country, state, or county	63	Classifying the collection	71
One mining district, mine, or locality	63	Cataloging	71
Single crystals	63	The ledger	72
Gem stone minerals	63	The card catalog	72
Pseudomorphs and unusual mineral forms	63	Numbering the collection	72
Showy or attractive minerals	63	Catalog data; how to list a specimen	74
Ores and related minerals	63	Labeling	75
Minerals personally collected	63	Types of labels	75
		How to label	76
Chapter 7 Housing, Cataloging, and Labeling	64	Numbering specimens; identification methods	77
Housing the collection	64	Chapter 8 Value of Mineral Specimens	78
Specimen or collection boxes	64	Determining factors	78
Open shelving	65	Abundance and rareness	78
Glass cases	65	Classification as a gem stone	79
Drawer cabinets	67	Over-all quality factors	79
Specimen trays	67	Market	80
Glass or plastic vials or boxes	68	Trading	80
Specimen bases	68	Price	80
Risers	68	A guide to buying specimens	80
Guides to good display techniques	68		

Bibliography

Most of the books listed below have been used as reference sources while preparing the manuscript for this book. Those prefaced by an asterisk are considered by the author as being particularly valuable in establishing a basic library in the field of mineralogy.

Berry, L. G., and Mason, B. *Mineralogy, concepts, descriptions, determinations.* San Francisco. W. H. Freeman & Co. 630 p.

Dake, Henry Carl (and others). *Quartz family minerals; a handbook for the mineral collector.* New York: McGraw-Hill Book Company, Inc., 1938, 304 p.

Dana, Edward Salisbury. *Minerals and how to study them;* revised by Cornelius S. Hurlbut, jr., third edition. New York: John Wiley & Sons, Inc., 1949, 332 p. (Also in paper back.)

*   . *A textbook of mineralogy with an extended treatise on crystallography and physical mineralogy;* fourth edition rev. and enl. by William E. Ford, New York: John Wiley & Sons, Inc., 1932, 851 p.

*Dana, James Dwight and E. S. Dana. *The System of mineralogy;* seventh edition entirely rewritten and greatly enlarged by Charles Palache (and others). New York: John Wiley & Sons, Inc., 1944-1951, 3v. (Vol. 4 and 5 in preparation.)

Dunbar, Carl Owen. *Historical geology.* New York: John Wiley & Sons, Inc., 1949, 567 p.

*English, George Letchworth and D. E. Jensen. *Getting acquainted with minerals;* revised edition. New York: John Wiley & Sons. Inc., 1958, 362 p.

Kraus, Edward Henry and C. B. Slawson. *Gems and gem minerals;* fifth edition. New York: McGraw-Hill Book Company, Inc., 1947, 332 p.

       (and others). *Mineralogy, an introduction to the study of minerals and crystals;* fifth edition. New York: McGraw-Hill Book Company, Inc. 1959, 686 p.

Longwell, Chester Ray (and others). *Physical geology;* third edition. New York: John Wiley & Sons, Inc., 1948, 602 p.

*Pough, Frederick Harvey. *A field guide to rocks and minerals,* second edition. (The Peterson field guide series.) Boston: Houghton Mifflin Company, 1955, 349 p.

Sinkankas, John. *Mineralogy for amateurs.* Princeton, N.J. D. Van Nostrand Co. 1964, 585 p.

       . *Gemstones and minerals, how and where to find them.* Princeton, N.J. D. Van Nostrand Co. 1961. 387 p.

Speckels, M. L. *The complete guide to micromounts.* Mentone, Calif. Gembooks. 1965. 97 p.

Zim, Herbert Spencer and P. R. Shaffer. *Rocks and minerals; a guide to familiar minerals, gems, ores and rocks,* illustrated by Raymond Perlman. (Golden nature guide.) New York: Golden Press, 1957, 160 p.

Chapter 1

Becoming Familiar With Minerals and Crystals

In a letter to this author, Dr. Clifford Frondell, Professor of Mineralogy and Curator of the Mineralogical Museum, Harvard University, said, "In my knowledge many great natural scientists started out not by reading books but by collecting things."

This is a book on the fundamentals of collecting minerals. It has been designed for the mineral hobbyist, both the beginning and adanced collectors as well as the interested student or teacher. It is not a book on how to identify rocks and minerals, a field guide to mineralogy, or one on lapidary arts and crafts. Instead, it places emphasis on those aspects of mineralogy that are most important to the study, enjoyment, and maintenance of a private collection. This means the how, what, and where of mineral and crystal collecting, the physical properties vocabulary of the hobby, as well as the different types of collections and specializations. Of equal importance are the organization, housing and cataloging of a mineral collection. Also included are dozens of professional tips that are the product of many years' experience of an avid collector of crystallized minerals.

Although this book is focused on minerals and crystals, the fossil and lapidary enthusiast will find the sections on collecting, organization, display, cataloging and housing of equal value to their fields of interest. This is a guide to the more-or-less established procedures of mineral collecting.

Archeology clearly indicates that man has been interested in precious and semi-precious stones for thousands of years. However, the organized hobby of collecting minerals and crystals is about three hundred years old, somewhat correlated with the beginning of scientific study of mineralogy. In the early stages of development, collecting minerals was primarily limited to university and college scientists who were seeking knowledge in the then budding fields of mineralogy and crystallography. In the early 1800's, good private collectons of minerals began to be assembled, along with major organized efforts on the part of colleges and universities to establish research and reference collections. By the 1850's many fine private collections of minerals existed in both Europe and the United States, and the avocation of collecting specimens was well established.

In the 1930's excellent home lapidary equipment capable of producng professional results became available. At this time the general field of mineral collecting gained many additional hobbyists, and at the same time boomed in the direction of the lapidary arts. Emphasis was on the cutting and polishing of semi-precious mineral materials such as agates and petrified wood, manufacturing various items of jewelry, cutting of spheres and the faceting of gems.

Lapidary is a voluminous subject and by itself is one of the fastest growing hobbies in the nation. Its popularity has more or less created a misconception that few people collect minerals and crystals. This is absolutely not the case since the number of mineral collectors has grown steadily, if not as dramatically, as lapidary.

Since about 1950 there has been a marked renewal in the mineral and crystal hobby with emphasis upon its scientific aspects plus interest in a "professional collection." What this means is simply an organized pattern of collecting and a systematic development around one of various collection types or formats. This has done much to promote the growth of many outstanding private mineral collections as well as to foster much greater understanding and interest in the science of mineralogy.

Instrumental in this growth, as well as in the lapidary arts and crafts, have been the numerous gem, mineral, and lapidary societies throughout the nation, especially through their affiliated state and national organizations.

Why collect minerals? Like any hobby, this is entirely a personal matter. Hob-

The seashore is an excellent source of specimens which may be found in adjacent cliffs and among exposed rocks on the beach.

bies cannot be forced or dictated. Unless you are a professional mineralogist, museum curator, or mineral dealer, you collect minerals because it is interesting, and provides a relaxing outlet for spare time. These factors would apply to many hobbies; however, there are a few somewhat unique aspects of the hobby of collecting minerals which make it particularly attractive. Some of these are as follows:

1. Perhaps the most important of all aspects of collecting minerals is that it is an outstanding *family* hobby, and it gets you out-of-doors. Hiking or camping along the seashore, in the foothills and high mountains, or in desert country, can be a part of this hobby. Not only does it have value as an avocation but is highly recreational in combining a well rounded program of activity between work at home and in the field. Weekend, holiday, or vacation trips can be devoted to, or planned to include, collecting specimens.

2. It is a hobby that can be followed to any degree a person desires in terms of study. This varies from the casual collector who enjoys interesting rocks and minerals to the advanced collector who becomes involved in the chemical analysis of minerals or the complexities of advanced crystallography. In other words, it is far from a limited hobby in terms of expansion and continuing study.

3. It is a stimulating scientific hobby, particularly for young people, since mineralogy is one of the basic physical sciences opening up the vastness of the associated fields of geology, chemistry, physics and paleontology.

4. There is a true uniqueness about each individual mineral specimen in a collection, for there never will be another exactly like it. Therefore, there is not the degree of commonness or repitition that is found in other collecting hobbies.

5. As a scientific hobby, there is always the possibility of making a new find or discovery of importance to the field, or specimen of outstanding beauty and, in some cases, considerable value. This lends excitement and anticipation to many field trips.

6. Minerals lend themselves to handsome, attractive, and interesting displays that can add much to a home.

7. The organization, cataloging, labeling, and displaying of a mineral collection in a professional manner is both satisfying and rewardng. It requires study and patience and can be a source of justifiable pride.

High atop a peak in the La Plata Mountains of southern Colorado, the author and students examine specimens from a huge outcropping of barite, quartz and celestite crystals.

8. A fine collection of minerals, properly labeled and catalogued, has very real monetary value.

In becoming familiar with minerals there is no substitute for observing and personally studying mineral specimens first hand in order to learn best their physical properties and to become proficient at recognizing and identifying them. Careful study and handling of specmens is necessary to reveal their nature. You cannot learn all about minerals just by looking at them or reading about them.

Perhaps the question most frequently asked of advanced collectors is "how do you learn or find out about minerals?" There is no single complete answer, but the collector gains his knowledge as a result of the following sources.

1. Publications
2. Museums
3. Mineralogical, gem, or lapidary clubs or societies
4. Private collectors and collections
5. Formal courses
6. Dealers
7. Field experience

PUBLICATIONS: There are a number of fine books in the field of geology and mineralogy which range from beginners' manuals and field-guides to advanced studies covering various aspects of geology, petrography, mineralogy, and crystallography. Most public libraries have such titles, but it is also wise for a collector to build his own library of reference books. A standard textbook or two on geology, an introductory book to mineralogy, a field-guide to mineral identification, and the volumes of Dana might comprise a good investment.

There are a few monthly journals and mineral magazines published in the mineral and lapidary hobby fields. Subscribing to one or more of these provides an interesting and current source of information and localities. Publications such as *Gems and Minerals*, *Rocks and Minerals*, *Earth Science*, and the *Lapidary Journal* are examples. For the more advanced collector, especially from the standpoint of mineralogy, the official organ of the Mineralogical Society of America, *The American Mineralogist*, published by the University of Michigan, is outstanding. A somewhat similar technical publication entitled *Economic Geology* frequently contains articles of interest to advanced collectors.

Regarding localities, there are a number of privately published guide booklets on a local or regionalized basis covering such areas as California, Nevada, Arizona, the Midwest and Colorado These are usually obtainable through mineral dealers, bookstores, or through advertisements appearing in the various mineral magazines. The number of localities covered by detailed published maps and data would keep a collector on the road for a good many months!

Another source of very valuable publications is in the various state university and college departments of geology, mineralogy or conservation. Each of the fifty states has somewhat different governmental control over mining and mineral resources, but the state or its educational institutions may publish various research papers, reports, pamphlets or information bulletins that are very useful to collectors. California, New Mexico, Arizona, and Colorado have all issued reports at times that prove of great interest to mineral collectors, not only from the standpoint of localities that may be discussed, but also because they contain an abundance of valuable and interesting information.

Similarly, the United States Geological Survey publications and reports on various mining areas may be of use, and the standard United States topographic maps published by the Survey are often valuable aids to effective field collecting. Topographic maps always indicate known mines.

MINERALOGICAL OR GEM SOCIETIES: There are several hundred active mineral or gem societies or clubs in the country and membership continues to grow. Joining a local mineral or gem society is certainly one of the very best ways to become familiar with minerals. The societies or clubs, have regular meetings at which various speakers present subjects of interest to hobbyists. In addition to regular meetings most societies participate in either their own annual show or combine with other societies on a state-wide or regional basis for such a show. Much can be learned from participating in these by studying other collectors' exhibits, noting especially those winning awards, and examining dealers' displays of material offered for sale.

One of the most important society activities is the field trip planned for collecting local and regional material. There is hardly a better way to find out how to collect specimens than by accompanying experienced collectors on field trips.

Joining a mineralogical society has another valuable attribute in that friendships are established with others who share the same hobby — with whom you can talk the same language. This also provides the opportunity to study and examine

A weekend encampment held by the California Federation of Mineralogical Societies, near Inyokern, in the Mojave Desert.

The Smithsonian Institution exhibits what is probably the finest collection of minerals in the world. This dramatic and beautiful display features crystals of the sulfate and borate types. Eighteen different types are shown. (Smithsonian Institution photo)

other collections. There is no better way to learn about minerals than to see and examine them.

COURSES: In most large communities or others where colleges or universities may be located, some sort of geology or mineralogy course may be available on the campus, offered by extension, or by the adult division of the public school system. If a hobbyist has had no formal education in geology or mineralogy, a beginning course in the subject is recommended. It provides useful background information. Sometimes basic introductory courses in mineralogy may be offered by some member of the mineralogical society, and many members are highly qualified in this regard.

MUSEUMS: Visiting public museums which feature exhibits of mineral specimens is one of the best ways to become familiar with the field. There are hundreds of local, state, college or university museums throughout the country which include some sort of mineral display. As might be imagined, some displays are much better in organization, quality of specimens, and display techniques than others.

When visiting a museum's mineral collection special attention should be given the following:

1. Allow sufficient time to see and enjoy the displays at your leisure.
2. Carefully read and study displays giving basic information regarding rocks, minerals, crystals, as well as local geology. Most museums have several such displays which each museum curator may present in a slightly different manner.
3. Pay particular attention to displays which feature or specialize in local or regional mineral specimens. Most museums have at least one or two such special exhibits and usually these are outstanding and cannot be equalled in other museums.
4. Note the variety of forms a given mineral may assume. Observing fine museum specimens is the best way to become familiar with this aspect of mineralogy.
5. Observe the over-all quality of specimens exhibited as well as the various display techniques used.
6. Give attention to both the mineral specimen itself and its locality. Familiarize yourself with localities, since these are often highly distinctive and important to the collector.

Following are some of the notable mineral exhibits in the United States and Canada. When planning an extended field trip or vacation, a half day or full day spent at these museums is well worth the time and effort.

1. The Smithsonian Institution (United States National Museum), in Washington, D.C., now contains what is generally accepted as the world's most outstanding collection and display of fine crystallized mineral specimens. The magnificent exhibit area of the Smithsonian, portions of which appear in the illustrations, was redesigned in 1959 and has proven an outstanding attraction even for those not particularly interested in minerals. No trip to Washington, D.C., should be planned without including this museum.

2. The American Museum of Natural History, New York City, N.Y., has long exhibited one of the finest displays of crystallized minerals in the nation. This famed museum contains many often-pictured specimens, especially in the gemstone field. The spacious museum halls include a number of instructional exhibits on rocks and minerals.

3. The Harvard University Mineralogical Museum, Cambridge, Massachusetts, houses one of the most complete mineral collections in the world. Hundreds of outstanding crystallized specimens, as well as very rare minerals, are on display. Harvard has been the center of the revision activities on the new volumes of *Dana's System of Mineralogy*.

Unusual arrangement of specimens and excellent lighting facilitate viewing this display of silicates at the Smithsonian Institution. Hobbyists too may greatly enhance the appearance of their collections by proper display. (Smithsonian Institution photo)

4. The California Academy of Sciences, San Francisco, California, now exhibits the finest specimens from the private collection of the late Mr. M. Vonson. Displayed in two sections of the Academy, Mr. Vonson specialized in large, showy, quality specimens, and the display areas include many informative exhibits on minerals and mineral formation. Also at the Academy is the gem-mineral and lapidary collection of Mr. William Pitts which is a notable collection of its type.

5. Chicago Natural History Museum, Chicago, Illinois. Hall. No. 35, of this great museum is devoted to cases of fine mineral specimens from all over the world. Included in the hall with the minerals is an exceptional display of meteorites. This museum was formerly called The Field Museum.

6. State of California, Department of Natural Resources, Division of Mines, Ferry Building, San Francisco, California. This extensive display features minerals of the State of California, but also includes specimens of similar types from other localities. It is a large reference collection.

7. Royal Ontario Museum of Mineralogy and Geology, Toronto, Canada. This excellent display features the minerals and geology of Canada, but contains specimens from many other localities as well. It is perhaps the most outstanding display of its type in Canada.

There are several other good collections, but those previously mentioned are the most outstanding and comprehensive. Smaller, sometimes specialized, fine collectons will also be seen at the following locations in the United States and Canada.

Arizona:
  Geology Department, University of Arizona, Tucson.
  Petrified Forest National Monument Headquarters, near Holbrook

California:
  California Institute of Technology, Geology Department, Pasadena
  Los Angeles County Museum, Los Angeles
  San Diego Natural History Museum, San Diego
  Santa Barbara Museum of National History, Santa Barbara
  University of California, Geology Department, Berkeley

Colorado:
  Colorado Bureau of Mines, Denver
  University of Colorado Museum, Boulder

Connecticut:
  Peabody Museum of Natural History, New Haven

Georgia:
  Georgia State Museum, Atlanta

Illinois:
  Museum of Natural History, University of Illinois, Urbana

Indiana:
  Indiana State Museum, Indianapolis

Kansas:
  University of Kansas Museum, Lawrence

Louisiana:
  Louisiana State Museum, New Orleans

Massachusetts:
  The Museum of Natural History, Springfield

Michigan:
  Cranbrook Institute of Science, Bloomfield Hills
  Michigan College of Mining and Technology, Houghton
  Mineralogy Museum, The University of Michigan, Ann Arbor

Minerals of the famed Tri-State Mining District (where Missouri, Kansas and Oklahoma meet) are featured in this display. The area has produced quantities of fine crystallized specimens for many years and continues prolific. (Smithsonian Institution photo)

Gem stones and gem stone material in rough and finished forms, are featured at Morgan Hall, American Museum of Natural History, New York. In the center case, above, is a six-inch rock crystal sphere. (Courtesy of American Museum of Natural History)

Missouri:
  Missouri Resources Museum, Jefferson City
  University of Missouri Museum, Columbia

Montana:
  Montana School of Mines, Butte

New Jersey:
  Paterson Museum, Paterson
  State Museum of New Jersey, Trenton

New Mexico:
  University of New Mexico Geology Museum, Albuquerque
  New Mexico Institute of Mining and Technology, Socorro

New York:
  Buffalo Museum of Science, Buffalo
  McGraw Hall, Cornell University, Ithaca
  New York State Museum, Albany

Ohio:
  Cleveland Museum of Natural History, Cleveland
  Ohio University Museum, Athens

Pennsylvania:
  Philadelphia Academy of Natural Sciences, Philadelphia
  The Carnegie Museum, Pittsburg

South Dakota:
  South Dakota School of Mines and Technology, Rapid City

Washington:
  Washington State Museum, University of Washington, Seattle

Canada:
  National Museum of Canada, Ottawa, Ontario
  Peter Redpath Museum, Montreal, Quebec

Section of the main display room at The California Academy of Sciences, San Francisco, showing specimens from the extensive M. Vonsen Collection.   (California Academy of Sciences photo)

These cases at Harvard's Mineralogical Museum feature exceptionally fine single crystals.
(Harvard University photo)

Chicago Natural History Museum's Hall 35 has fine display of minerals, meteorites.
(Chicago Natural History Museum photo)

Chapter 2

Physical Properties Of Minerals and Crystals

This is an introduction to the basic physical factors and characteristics of rocks, minerals, and crystals. These are the essential concepts and factors that allow the collector to understand and scientifically describe his specimens. This involves the specialized terms constituting the language of mineralogy as applied to the hobby of collecting crystallized minerals.

Minerals have certain identifying properties, which cause them to differentiate one from the other, that are determined by four basic methods: (1) chemical analysis; (2) X-ray analysis; (3) optical properties; and (4) observation of physical properties.

The physical and some optical properties of minerals are the most important to collectors in terms of identifying, describing, and enjoying specimens as well as maintaining and displaying them in a collection. These physical properties of minerals—crystals, color, streak, luster, optical factors, hardness, cleavage, fracture, specific gravity, tenacity, and structure, all are *variable*. Many minerals occur in a well-known but astonishingly wide variation, and combinations of some of these properties as individual examples are found throughout the world at different localities. It becomes just as important for a collector to be familiar with the span of variation of a given mineral as it is to be well versed in basic physical properties. The physical variations of minerals add much more depth to the study and hobby than might appear. Unusual and rare variations or combinations are often choice collectors' items.

The physical properties of minerals are the most easily seen in a specimen. You cannot see a mineral's atomic or chemical structure, only an external manifestation of it. You can, however, readily observe factors such as color, crystal shape, structural form, optical qualities; and with simple tests you can estimate hardness and specific gravity. For most collectors, this, plus looking up the chemical formula is sufficient. From the standpoint of the advanced hobbyist, the professional mineralogist or mineral dealer, minerals are almost always described both in terms of their outstanding physical characteristics (the general appearance and structural form of the specimen) and chemical composition. Since this book is not a listing of the over four thousand different mineral species, emphasis is placed upon outlining important physical and some optical properties.

The *chemical composition* of minerals is of prime interest to most mineral collectors. In the science of mineralogy, minerals are grouped and classified according to chemistry, and minerals are the product of natural chemistry. However, determining the chemical composition of a mineral or mineral intergrowth is difficult or impossible for most collectors and is a specialized field. Taking an introductory course in determinative mineralogy at some college or perhaps evening adult school will provide the collectors with a number of simple chemical and blow-pipe tests facilitating the determination of most common minerals. Beyond this the field becomes fairly complicated but increasingly interesting.

Consequently, collectors interested in the chemical composition of minerals refer to some standard textbook, field guide, or mineral listing such as the various volumes of *Dana's System of Mineralogy*, or Pough's *Field Guide to Rocks and Minerals,* or similar publications which vary in terms of coverage and complexity. These provide the laboratory analysis of the chemical composition of known mineral species.

THE DIFFERENCE BETWEEN A ROCK AND A MINERAL: All rocks and minerals are composed of chemical compounds of elements, so this is where to begin. An *element* is a substance which cannot be further separated into unlike parts by ordinary types of chemical change, or be made by chemical union. There are 102 known

Fig. 2-1. Left to right above are specimens of a type of granite and the 3 minerals (quartz, feldspar, mica) of which it is composed.

elements and only some 20 are found in a pure natural state.

A *mineral* is any chemical element or compound occuring as a natural substance as a part of the earth's crust, having a varying but characteristic chemical composition and, with few exceptions, an orderly atomic structure sometimes expressed by the external form of the mineral as a crystal.

Most *rocks* are composed of aggregates or combinations of one or more minerals. Another way of saying this is that minerals are usually distinctive individual substances while rocks are physical mixtures of minerals.

In Fig. 2-1 this difference is clearly shown by using as an example one type of the common rock called granite. Granite is a rock frequently composed of three minerals—quartz, feldspar (orthoclase or microcline) and biotite mica. When these three minerals occur in nature mixed together as a part of the earth's crust they can form the rock granite. This is a very obvious example, but establishes the basic difference between rocks and minerals. Another common granite type involves hornblend or augite instead of mica.

THE PHYSICAL PROPERTIES OF MINERALS: In the following pages, only those physical properties of a mineral which may be readily observed by the eye and touch, or lie within the scope of simple tests, are included. Due to the depth of the

Figs. 2-2, 2-3. At left is a fine sample of well-crystallized mineral specimen of gypsum, variety selenite, from Cave of Swords, Naica, Chihuahua, Mexico, from the M. Vonsen Collection at The California Academy of Sciences. At right is cleaved rhomb of water-clear calcite showing interesting optical phenomenon of double refraction, from Convict Lake in the Sierra Nevada, Mono County, California.

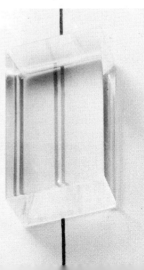

subject and its complexity, no chemical procedures are included, nor are seldom-used terms.

CRYSTALS: No other physical property of a mineral is more important or highly distinctive than a natural, well crystallized example as in Fig. 2-2. Crystals facilitate identification of a mineral since two mineral species seldom crystallize exactly alike. Given a well-developed crystal of a mineral, in all probability identification can be made from observation of the crystal shape alone. From the standpoint of the collector they are the "flowers" of the mineral kingdom and the most sought after type of specimen. The following chapter is entirely devoted to crystals, crystallization, and other forms which are interesting from the collector's standpoint.

COLOR: Color is perhaps the most readily determinable physical property of a mineral, yet at times it can be very misleading in terms of identification. Minerals are likely to occur in more than one characteristic color. Those that are quite consistent in appearing in one color, or closely related shades, are called "idiochromatic." Typical examples would be malachite — green, azurite — blue, and sulfur — yellow, although in the last, its color ranges through a wide variety of shades from orange yellow to pale lemon yellow.

A great many minerals occur in a wide variety of colors due to the presence (and percentage) of chemical impurities or microscopic inclusions of another mineral. Such minerals are called "allochromatic" and may occur as individual specimens of various colors or as a single specimen showing a stratification or "zoning" of colors. Quartz, for example, is most characteristically pure white to colorless; however, it is found in purple crystals called amethyst, brown to black crystals called smoky quartz, and a golden yellow known as citrine. Many other colors of quartz are less common. Another example would be the mineral garnet. Garnet is almost always associated with the color red by the layman; however, fine garnet crystal specimens are found in lavender, hyacinth, pink, white, yellow, purple, cinnamon, olive, brown, black, and a spectacular emerald green.

Other minerals occurring in a wide variety of colors are barite, calcite, tourmaline, and fluorite, which are often strongly zoned within a single crystal. Fluorite from the noted locality of Rosiclare, Hardin County, Illinois, has been found with huge transparent cubes not only liberally sprinkled with marcasite crystal inclusions but strongly zoned in bands of white, lavender, gold and dark purple! The most striking example of mineral color zoning is found in gemstone tourmaline crystals, particularly those from the pegmatite dikes of San Diego County, California, and Exford County, Maine. Such crystals are often bi-colored, or even tri-colored, with zoned areas of green, red, pink and colorless. Crystals showing two colors are referred to as "bi-colored," and if three colors appear, "tri-colored."

The specific color variations of a mineral frequently carry a sub-species or variety name while belonging to a specific mineral family. For example, the mentioned emerald green garnet is a calcium-chromium garnet called uvarovite. The pinkish-red tourmalines are called rubellite, the blue ones indicolite, and opaque coal-black are the variety schorl.

So far only crystals have been mentioned, but the massive or crystalline form of a mineral can show color variations as well. An example would be smithsonite, a zinc-carbonate mineral not commonly found well-crystallized, but more frequently in massive type specimens of the mammillary and reniform structural forms. Smithsonite really has no particular characteristic color. It is found in white, a avriety of greys, black, brown, yellow and a number of beautiful shades of green and blue.

The most important thing for the collector to remember about color in minerals is to be aware of the *variety* of colors a mineral can assume, and then keep an eye out for odd and unusual colors or color combinations. Also, in building a collection, one should try to obtain as many different colors of a given mineral as opportunity permits. Such common mineral species, such as calcite or quartz, but of an unusual color or color combinations, can be valuable additions to any collection, private or museum.

If a specimen shows a changeable silky sheen when moved about under a light source, this is called "chatoyancy" and is characteristic of many minerals having fibrous structure such as the satin-spar variety of gypsum, crocidolite (tigereye), chrysoberyl (cat's-eye), and some tourmalines. "Tarnish" is a color term that is self-explanatory and refers to the characteristic of some metallic minerals to oxidize upon exposure to the air, such as native copper, pyrite, and marcasite. "Color play" or "schiller" refers to a few minerals showing a remarkable change or play of colors when turned a different angles. The two feldspar family species, moonstone (true moonstone is a feldspar variety called albite), and labradorite, both show schiller in terms of an opalescent pearly-blue and a shimmering iridescent peacock play of colors. The color play readily observed in precious opals is referred to as "fire," and in non-gem quality materials as "opalescence."

STREAK: "Streak" is another physical property used in mineral determination that is primarily based on color. Streak is the color of the powdered mineral as observed by a scratch against a fine abrasive surface. Minerals may vary in apparent exterior color but the powdered streak color remains fairly constant.

It is a simple matter to perform a streak test. The "streak" is determined by drawing a corner of the specimen over the surface of a piece of unglazed porcelain such as the back side of a kitchen or bathroom tile. This scratch leaves a distinctive colored streak. The streak of a mineral is given in almost all texts and mineral species listings.

Many different minerals may appear the same and the difference is quickly resolved by the streak test. For example, hematite and magnetite, both iron minerals, may appear identical in certain specimens; however, hematite yields a reddish brown streak and magnetite one that is black. Not all minerals will leave a streak since those of a hardness of 6 and above may be too hard to be abraded by the unglazed porcelain surface.

It is interesting to note that streak is generally a more reliable identification and diagnostic physical property than a mineral's observed and apparent color.

PHOTOSENSITIVITY AND FADING: It is not uncommon for certain minerals to show a change in color after a period of time, due to some type of oxidation on the exterior surface or simply by fading. A few silver minerals, such as proustite and pyrargyrite are extremely sensitive to light and are called *photosensitive*. These minerals, often referred to as ruby silver, occur in fine rich ruby red crystals, which upon prolonged exposure assume a dull grayish color. This can be removed by gentle brushing with lukewarm water and mild soap. Many minerals assume a dull luster, or even a crust as is the case with several borates, as they dehydrate or hydrate upon exposure to air. In addition, some crystallized mineral specimens simply *fade*. This is particularly true of golden calcite which will fade to a pale lemon yellow in some specimens as the years pass.

The mineral hackmanite exhibits what is called *reversible photosensitivity*. This is a reddish violet to pink mineral, which fades to pure white upon exposure to light. However, upon exposing the mineral to ultraviolet light radiation, the original color can be restored, though it will again fade slowly in normal light.

LUSTRE: The term "luster" refers to the general manner and natural appearance of a mineral surface as it reflects light when viewed under a strong light source. Essentially, mineral luster falls in two main divisions, metallic and non-metallic. Following are the more-or-less standardized luster terms:

Metallic: This is the characteristic luster of metals and refers to minerals having an obvious metal-like appearance to their natural surface. If the metallic luster is uneven, mixed, or imperfect it is called *sub-metallic*.

The remaining terms are all applicable to minerals in the non-metallic luster division.

Vitreous: Appearing glassy, or glass-like.

Adamantine: A brilliant, highly lustrous surface, like a diamond.

Resinous: Appears like resin, wax-like.

Greasy: Appears to have been oiled.

Pearly: Like mother-of-pearl, and common among minerals of laminated structure.

Silky: Self-explanatory in appearance, usually confined to minerals of fibrous structure.

Dull or Earthy: Lacking any brilliance; lusterless by comparison.

There are other descriptive terms applicable to luster, but those mentioned are most commonly used. Sometimes the degree of luster is also described and the terms are *dull, splendent, shining,* and *glistening*.

Like other physcal properties, luster is variable and a given mineral from different localities occurring in different structural forms will have varying lusters. Also, an individual specimen may well show more than one type of luster. An apophyllite crystal may be dull and pearly at the base, but terminate in a transparent vitreous termination. Such situations are usually stated in terms of range, such as "a luster ranging from pearly to vitreous" or whatever the relationship may be.

MINERAL OPTICS: The degree to which light passes or does not pass through a mineral is called *diaphaneity*. The terms used to describe this condition are as follows.

If a mineral passes absolutely no light whatsoever, even on thin edges, it is said to be *opaque*. If a specimen is clear and easily seen through, the term is *transparent*. If the mineral passes some light but only a dim outline of objects may be seen through it, it is *translucent*. If it passes light only on thin edges or through a tiny sliver-like fragment, it is *semi-translucent, or sub-translucent*.

Some minerals may appear opaque at first glance but translucent on thin edges when carefully observed. Some specimens of rutile appear, on the surface, opaque and a silvery-metallic or black. When examined on thin edges or against a strong light source, they may be found to be of a deep rich ruby red color!

INDEX OF REFRACTION, OR REFRACTIVE INDEX: This is a highly technical physical property and one seldom cited by collectors, unless specializing in gemstone minerals. Basically, the refraction of a mineral refers to the fact that the rays of light entering a given translucent or transparent mineral are always bent to exactly the same degree every time. All minerals differ somewhat regarding the degree of bending, so this becomes another method of mineral identification. However, it requires considerable study as well as specialized equipment. Diamond has a high index of refraction, meaning that a ray of light will be refracted, or bounced around, within the crystal many times before escaping. This accounts for the famed brilliance of diamonds, and when cut, the facets are so placed on the stone as to take advantage of the high index of refraction factor.

DOUBLE REFRACTION: A common optical property of many transparent minerals, especially calcite, is ability to split light in such a manner as to produce a double image. In Fig. 2-3 a water-clear transparent cleaved rhombohedron of calcite from Convict Lake, Mono County, California, demonstrates this interesting phenomenon. Double refraction occurs to varying degrees in certain minerals of five of the six crystal systems, but occurs only under very unusual conditions within the cubic system.

ASTERISM: A few minerals show the unique quality called asterism. This is a star-like radiation from within the mineral when viewed either by reflected or transmitted light. Some phlogopite mica shows a pronounced star when viewed against a light source. The precious gemstone, star sapphire, is a sapphire that shows asterism by reflected light, as do star garnets and some quartz.

HARDNESS: Minerals differ considerably in hardness. The relative hardness of a mineral is usually determined by scratching the unknown mineral with a mineral of a known hardness. By this method an unknown is soon bracketed by a mineral that will scratch it and one that won't, thereby approximating its relative hardness. This is usually done according to a "scale of hardness" such as the well-known Mohs Scale, a seres of minerals listed from 1 to 10 in hardness — 1, talc, is the softest, and 10, diamond, the hardest.

Mohs Scale of Hardness

1. Talc
2. Gypsum
3. Calcite
4. Fluorite
5. Apatite
6. Feldspar
7. Quartz
8. Topaz
9. Corundum
10. Diamond

The hardness of many minerals, when a variety of samples from a number of different localities are tested, will show a certain degree of variation. In other words a mineral as a species may vary in hardness. For example, the mineral chalcocite varies in hardness from $2\frac{1}{2}$ to 3, and would be expressed as H. $2\frac{1}{2}$-3; pyrite - H. 6-$6\frac{1}{2}$; samarskite - H. 5-6; sulfur - H. $1\frac{1}{2}$-$2\frac{1}{2}$. Hardness must always be tested on a fresh surface.

SPECIFIC GRAVITY: In mineralogy, the term specific gravity simply refers to the weight, or, more accurately, the relative density of a given mineral compared with an equal volume of water. The water factor has been established at a temperature of 4°C. It is a measure of ratio—*i.e.*, the weight of the specimen divided by the weight of an equal volume of water. Gold, for example, has a specific gravity of 19.3. In other words, a cubic inch of gold weighs 19.3 times as much as a cubic inch of water.

In the laboratory, specific gravity may be obtained by a chemical, beam or Jolly balance, or a device called a pycnometer. Such equipment is seldom available to the hobbyist, and the established specific gravity of minerals is given in almost all mineralogy books. The collector usually gives an estimate of the weight of a specimen when held in the hand. The following estimate terms are used frequently: *very light, light, medium heavy, heavy, very heavy.*

Specific gravity is a very important physical aspect of minerals (as long as the specimen is not intergrown) and the collector soon learns not to trust his eyes alone, but to count on "heft" or his sense of touch as well. An example in Fig. 2-4 shows two specimens that appear quite alike. They have the same luster, apparent cleavage, rough edges, transparency, general shape and color. However, one is barite with a gravity or G of 4.5 and the other is gypsum with a G of 2.3; one is almost twice as heavy as the other. This basic difference cannot be observed by the eye, but is immediately apparent by handling both specimens. Cerussite, a lead carbonate, is another mineral which can look surprisingly like barite or gypsum in some specimens, and it has a G of 6.5. There are dozens of possible "look-a-likes" in the mineral field.

CLEAVAGE AND PARTING: The tendency of certain minerals to break or cleave in definite directions or planes is called *cleavage*. Not all minerals show cleavage, but when it is present it is a distinctive physical property, with some minerals having two or more different cleavage planes. The so-called planes of cleavage are always planes of the mineral's inherent crystal structure. If a mineral cleaves with ease, showing a clean smooth surface, the cleavage is called *perfect* or *distinct*. If,

Figs. 2-4, 2-5. Minerals often resemble one another in several physical aspects. At top left is barite, with gypsum below it. The barite is twice as heavy and hard as the gypsum. The cleaved octahedrons of fluorite at right are man-made cleavages, though fluorite can and does sometimes occur naturally crystallized in this shape.

however, cleavage is difficult and yields an inferior, indistinct and rough surface, the cleavage is called *uneven* or *imperfect*.

It is important for the mineral collector to realize that cleavages are *not* crystals. Cleavages are the product of natural breakage or purposeful cleaving. A typical example of an often mistaken classification is seen in Fig. 2-5. This illustration shows several apparent "crystals" of the mineral fluorite; however, they are not natural crystals, but cleavages of this mineral. Fluorite has a highly distinctive cleavage pattern of a perfect octahedron, such as those illustrated. Wherever fluorite is found on the face of this earth it will cleave into this form. Fluorite does occur in natural crystals in the octahedral form, but it is infrequent. The most common natural crystallization is in a cube.

The term *parting* refers to another aspect of mineral breakage which constitutes a separation along a plane that is not necessarily related to the crystal structure. Parting is usually along some area of weakness that may be caused by pressure on the mineral, inclusions, or unusual twinning.

FRACTURE: When the direction of a break in a mineral specimen is not definite, such as with a cleavage plane, the mineral is said to *fracture*. The fractured surfaces appear differently in minerals and are defined as: *even, uneven, hackly, rough, splintery, earthy,* or *conchoidal*. These fracture terms are self explanatory except for the latter, conchoidal. This refers to a circular and smooth concave or convex fracture. Obsidian has a characteristic conchoidal fracture.

TENACITY: This defines how a mineral holds together or how it stands up under blows. If it can be flattened by hammering, it is *malleable*. If a specimen can be cut in sheet-like forms by a knife blade, it is *sectile*. It is *brittle* if it breaks or shatters when hammered, and *tough* if it generally resists breakage.

STRUCTURE: Most minerals do not commonly occur as single, perfect, individual crystals, but rather as aggregates and groups of imperfectly-formed crystals, or as crystalline masses. Such grouping manifests itself in a variety of different mineral structural forms. Following, listed alphabetically, are the most common

Figs. 2-6, 2-7. At left are acicular, needle-like crystals of scolecite found at Teigarhorn, Berufjord, Iceland. The specimen at right is an excellent example of arborescent, tree-like structure, in a sample of desclozite from Abenab in South West Africa.

types of structural forms encountered by the mineral collector. These are the language of the field or the common descriptive terms of the hobby. Keep in mind that it is quite possible for a mineral to occur in a great variety of different structural forms, especially from different localities:

Acicular: Crystals in long, thin needle-like form such as in Fig. 2-6.

Aggregate: A crystal cluster, group, or mass of one or more minerals.

Arborescent: Branching tree-like aggregates of crystals, as in Fig. 2-7. A form characteristic of many native copper specimens, crystals occurring closely packed, forming sheets and curved plates.

Bladed: Blade-like, flattened and elongated crystals. One of the common crystal forms of barite. See Fig. 2-8.

Figs. 2-8, 2-9. The bladed crystal structure at left is a form often seen in barite. Specimen shown is from the New Glencrieff Mine at Wanlockhead, Dumfrisshire, Scotland. Botryoidal structure is illustrated at right, in azurite from Copper Queen Mine, Bisbee, Arizona.

 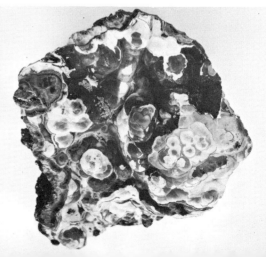

Figs. 2-10, 2-11. Coralloidal "flos ferri" form is shown at left in an aragonite specimen from Styria, Austria (Minerals Unlimited photo). At right is sample of manganese dendrites in shale, not fossils as might appear, from a locality in the Barstow, California, region.

Botryoidal: Compact mass of small rounded surfaces, similar to a bunch of grapes or bubbles. See Fig. 2-9. A similar, but smaller basic form than either mammillary or reniform structures.

Capillary: Finely hairlike. Finer and more delicate than acicular form.

Columnar: Composed of rod-like forms or columns, usually in parallel groups.

Concentric: A spherical or banded form, layer upon layer about a common center. Shell like.

Coralloidal: Coral-like branching and twisting forms. Examples are commonly, found in aragonite, when it occurs as helictites in caves, or as the variety called "flos ferri" shown in Fig. 2-10.

Concretionary: A rounded nodular mass originating in sedimentary deposits. A few minerals occur as concretions, but the form is also important when geodal and lined with crystals.

Crystal: A natural mineral substance bounded by natural plane surfaces or faces.

Crystalline: A homogenous mineral mass of compact closely knit crystals, varying in texture from coarse to fine. Possesses crystal structure in a compact form with or without specific crystal form.

Cryptocrystalline: A crystalline form in which the structure is so fine it cannot be seen with the naked eye.

Dendritic: Fernlike in form. A branching, extremely delicate variation of arborescent structure. Often mistaken for fossil remains, as clearly illustrated in Fig. 2-11.

Druse or drusy: A thin encrustation or coating of crystals on a rock surface.

Etched: Crystals are encountered which have been highly modified or whose crystal faces have been pitted or scarred deeply, the product of true etching of the mineral by secondary action. Etching often assumes some sort of a definite pattern on a crystal face.

Fibrous: Fiber-like, or fibrous appearing, as in Fig. 2-12 showing asbestos. Fibers may be radiating, parallel, or divergent.

 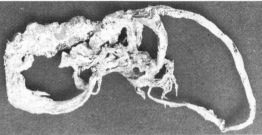

Figs. 2-12, 2-13. Specimen of natural, unprocessed asbestos at left is good example of fibrous structure. The filiform, lace or wire-like form, along with heavier "ropes" and crystals, is illustrated at right, in a prized specimen of native silver found at Kongsberg, Norway, and now in the private collection of Robert O. Deidrick.

Filiform: Thread or wire-like, often twisted and bent, like a filament. Natural wire-silver, (Fig 2-13), and wire-gold often occur in this structure.

Foliated: Composed of thin, separable leaves or laminae.

Geode or geodal: A hollow crystal-lined, rounded or spherical form is a geode, as in Fig. 2-14. Geodal refers to hollow crystal-lined cavities within a specimen.

Globular or *spherical*: Sphere-like forms. See Fig. 2-15.

Helictite: Refers to the more rare of cavern formations — that which protrudes in a branching manner from the sides of cavern rooms, or the sides of stalactites, as in Fig. 2-16.

Lenticular: Shaped like a lens.

Mammillary or *mammillated*: Large contiguous rounded surfaces; larger basic form than either botryoidal or reniform structure.

Massive: A solid mass of a given substance. May be crystalline internally but lacks any outward appearance of crystal structure or form.

Matrix: Refers to the basic host rock upon which crystals are located, or the rock surrounding a crystal specimen.

Micaceous: Composed of thin platy sheets, resembling mica.

Nodule or nodular: Rounded spherical forms varying from regular to irregular shapes.

Oolitic or pisolitic: Oolitic refers to a mass of minute rounded forms appear-

Figs. 2-14, 2-15. At left, geodal structure is apparent in quartz crystal geode from well-known geode area near Keokuk, Iowa. The small black flecks are millerite crystals. Globular form is shown at right in prehnite specimens from West Paterson, New Jersey, and in spherical aggregate sample of adamite from Mapimi, Durango, Mexico.

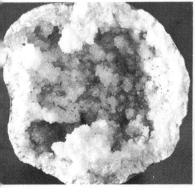 

Figs. 2-16, 2-17. Helectites of aragonite are shown at left, growing on sides of stalactites in Coral Gardens section of Timpanogos Cave National Monument, Utah. At right, reniform or kidney-like form seen in hematite "kidney ore" from Ulverstone, Lancashire, England.

ing like fish eggs. When the same form is on a larger scale approximating pea size it is termed pisolitic.

Plumose: A compact mass of plume-like, feathery, branching crystal growths.

Radiated or radial: A circular form spreading outward from a center, usually flattened. Characteristic of certain localities for wavellite, pectolite and disk-like forms of marcasite. Another use of the term refers to a group of crystals that seemingly burst in a radiating spike-like display from a common center. Black tourmaline, stibnite, stilbite, and quartz frequently occur in this structure.

Reniform: Rounded surfaces that appear kidney-like as in Fig. 2-17. Basic form larger than botryoidal and smaller than mammillary or mammillated.

Reticulated: An intricate intergrowth of crystals forming an entangled net-like form or structure. If the form shows knee-like bent forms it is called *geniculated*. Reticulation is found frequently in the mineral cerussite (Fig. 2-18).

Rosettes: A few minerals occur in a complex multiple crystal simulating a

Figs. 2-18, 2-19. Reticulated structure, left, is shown by the intersecting lattice-like network of crystals in specimen of cerussite from Mammoth St. Anthony Mine, Tiger, Arizona. At right is good example of hematite rosette, often called "iron roses," with adularia crystals. Specimen was found near Fibia, St. Gotthard, Switzerland.

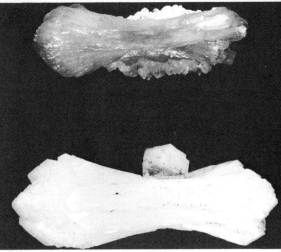

Figs. 2-20, 2-21. At left, quartz crystal from Tavetsch, Canton Grisons, Switzerland, is heavily rutilated with rutile crystals. Samples of stilbite at right demonstrate sheaf-like form most commonly found in this mineral. The form is apparent in these specimens. Top, from West Paterson, N.J.; bottom, from Teigarhorn, Berufjord, Iceland.

rose-like form. Common in barite, gypsum, and less frequently in hematite (Fig. 2-19).

Rutilated: Characterized by the inclusion of needles of the mineral rutile, as in rutilated quartz (Fig. 2-20).

Sheaf-like: A composite grouping of crystals appearing like a tied sheaf of wheat. Fine specimens of stilbite occur in this form, as in Fig. 2-21.

Stellate: A radiating structure which produces star-like forms. The mineral water, crystallizing in snowflakes, always assumes a star-like stellate structure.

Stalactitic: Formed as a hanging, pointed pendant resembling an icicle — like a stalactite. See Fig. 2-22.

Stalagmite: Essentially a limestone cavern formation, always rising from the floor of a cave toward the roof.

Striations: Minute straight line-like ridges or linear depressions on crystal

Figs. 2-22, 2-23. Stalactitic forms are seen at left, in malachite from Apex Mine near St. George, Utah; and in marcasite from Tri-State Mining District, Joplin, Missouri. At right are striated crystal faces, on a pyrite cube from the Flux Mine, Washington Camp, Arizona.

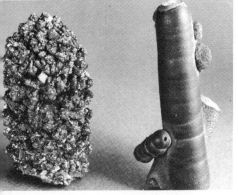

25

faces occurring during crystallization. Pyrite and tourmaline are often "striated." See Fig. 2-23. If the striations are very deep and grooved, the specimens are referred to as being "furrowed."

It should again be emphasized that these are basic structure forms, and that a given mineral species may occur in a number of these forms as found at different localities, or at a single locality. For example, the green copper carbonate, malachite, is found occasionally as crystals and commonly as massive specimens with mammillary and reniform surfaces. It can also occur in globular, nodular, geodal, dendritic, acicular, concentric and stalactitic forms, as well as fine capillary inclusions in calcite and quartz.

Some particular structure form of a mineral may be quite indicative of a particular locality, such as the reniform and mammillary hematite specimens from England, the hematite roses from Switzerland, or the fine single crystals of hematite from the Island of Elba.

The structural form of a mineral, as far as the mineral collector is concerned, can be equally important as the mineral itself and the locality involved.

ELECTRICITY: A few minerals, particularly tourmaline and some quartz, may exhibit both *pyroelectric* or *piezoelectric* characteristics. Pyroelectric refers to the property of a mineral gaining a minute electrical charge as it is either heated or cooled. The term piezoelectric means that the specimen becomes charged when pressure is applied. In both cases, a specimen will show polarity, with one end of a crystal positive, the other negative. A classic and simple experiment involves the dusting of a warm crystal of quartz or tourmaline with a mixture of sulfur and red lead powder. The powder should be blown onto the specimen through a screen of fine cloth to statically charge the powder particles. A rubber battery bulb or syringe is satisfactory. The sulfur will gather on the positive end of the specimen, the red lead on the negative. Topaz, sulfur, and a few other minerals may develop a slight electrical charge when rubbed briskly.

MAGNETISM: There are two types of magnetic qualities of interest to the collector. First, minerals that are magnetic; *i.e.,* they are attracted to a magnet. The second is a mineral that is a natural magnet by itself and will attract iron-bearing filings.

There are not too many minerals that are magnetic, with magnetite leading the list in terms of obvious attraction to a magnet. The minerals hematite, franklinite, some limonite, ilmenite, chromite, platinum and pyrrhotite are all somewhat magnetic, but the response can be very weak.

Nature's natural magnet is not too common and is a variety of magnetite referred to as "lodestone." This mineral can exhibit very strong magnetism and marked polarity. Specimens have been found with sufficient magnetic pull to hold a number of large nails, keys or similar objects to their surface.

FLUORESCENCE AND PHOSPHORESCENCE: These are related properties observed in certain minerals from certain localities. *Fluorescence* is the property of a mineral to emit light, other than its apparent natural color, during exposure to ultraviolet light radiation. *Phosphorescence* is much less common than fluorescence, and is the property of a mineral to continue to emit light, or glow, after the ultraviolet light source has been removed. Phosphorescence may be of a few seconds duration or may last many minutes.

Although fluorescence and phosphorescence of minerals are generally synonymous with ultraviolet rays, the same phenomena may be produced in certain specimens by a high electrical charge in a vacuum tube, by X-rays, and other rays connected with atomic energy research.

The actual cause of fluorescence and phosphorescence is a major venture into physics and mathematics, and a great deal about the phenomena remains unknown.

However, a basic (but technically over-simplified) explanation is as follows:

All minerals are composed of atoms, and these are composed of smaller particles called electrons revolving in orbits around the center or nucleus of the atom. When certain minerals are contacted by ultraviolet light, it upsets or disrupts the normal pattern of atomic particle arrangement. You might say that the particles are knocked out of orbit. During this disturbance the displaced particles give off energy in the form of heat, which in turn is observed as light. The various wavelengths of the light determine the color of the fluorescence. Most fluorescent substances return to normal stability immediately after the light source is removed; however, if the particles take a longer period of time to reorient themselves after the light source is turned off, the property of phosphorescence is seen.

Fluorescence and phospehorescence are best observed under ultraviolet light radiation rated at 2500 Angstrom Units. Most minerals fluoresce because of delicate balance between the mineral involved and some impurity which seemingly acts as a catalyst does in chemistry — it starts or triggers the action. For example, the mines around Franklin Furnace, New Jersey, have long been noted for their spectacular fluorescent specimens, perhaps the finest in the world. Of particular interest to this point is the brilliant red-fluorescing massive calcite from this locality. It fluoresces due to the presence of between 1% and 5% manganese, as an impurity, which activates the action. Calcite from this locality containing 3% and $3\frac{1}{2}$% manganese impurity fluoresces the most brilliant cerise red. If the specimens from this locality have less than 1% or more than 5% manganese there is no fluorescence. Also, not all calcite will fluoresce — not even all calcite from a given locality where fluorescence is known to be present.

Another example would be the mineral willemite, a brilliant green-fluorescing and phosphorescing mineral. It is strongly fluorescent from some localities, notably Franklin Furnace, and completely "cold" from others.

The term fluorescence comes from fluorite, a mineral that frequently fluoresces a fine blue, bluish-white or green. But again, not all fluorite will fluoresce.

Collecting fluorescent and phospherescent minerals constitutes the specialized hobby of some collectors, and a point of interest in terms of a suite of specimens for others. From the standpoint of crystallized minerals and showy specimens, the overwhelming majority of fluorescent minerals are very unattractive until the ultraviolet light is turned on. Most specimens are very dull, appearing as rock-like massive forms or aggregates. However, the vivid and intense colors under ultraviolet rays are one of the most dramatic factors in the entire field of mineralogy.

A common misconception among laymen is that fine, showy, crystallized mineral specimens will all turn another color under ultraviolet light. This is, indeed, very seldom the case, and with the exception of fluorite, autunite, and some calcite, would be limited to a few specimens.

The mineral scheelite, a valuable source of tungsten, is usually strongly fluorescent in yellow to bluish hues. Scheelite is difficult to distinguish in many ore bodies by natural observation; however, its fluorescent qualities are so outstanding that it is prospected for with portable ultraviolet lamps. Ore bodies are mined according to where the ultraviolet bloodhound indicates the scheelite to be.

LUMINESCENCE: This phenomenon is not to be confused with fluorescence or phosphorescence. *Triboluminescence* is the unusual feature shown by some minerals, particularly sphalerite and corundum from certain localities, of showing an internal spark-like flashing of light when scratched by a sharp metal point. The effect is best observed in darkness, and not all sphalerite or corundum will demonstrate triboluminescence.

The effect referred to as thermoluminescence, as is indicated by the word, refers to the property of a mineral to emit light or glow when low heat is applied to the specimen. It, too, is best observed in darkness, and fluorite and calcite frequently demonstrate this property.

RADIOACTIVITY: The phenomenon of radioactivity involves the atomic breakdown of a mineral, and for all practical purposes is confined to the various uranium minerals and related species. Some radioactive minerals, such as autunite, are fluorescent; but this is not a particularly distinguishing feature. Their unique property is apparent through the use of a Geiger counter, which picks up the invisible rays constantly emitted by the specimens and transfers them to sound waves, or by the ability of the mineral to take a picture of itself!

Many radioactive minerals will take their own picture if left on a photographic plate or sealed sheet of film for a period of 24 hours or more. The invisible rays penetrate and expose the photographic emulsion. The film is then developed in the normal manner and a regular print made from the negative. The resulting picture is called an "autoradiograph."

Despite the intensity with which some "hot" specimens, such as autunite, carnotite, and uraninite, activate a Geiger counter, radioactive minerals occurring in nature are considered harmless. Although technically emitting atomic radiation, the rays involved are of a type and intensity that is safe for collection specimens. The only damage they can do around the home is to ruin photographic film if inadvertently placed beside it, either in or out of a camera.

Figs. 3-1, 3-2. Despite apparent differences in size, shape and color, the angle between any two faces is the same on specimens of quartz crystals at left, from Arizona, Brazil, California, France, Italy, Japan, Nevada, North Carolina and Switzerland. At right is a classic example of the "fishtail" contact twin of calcite, found at Treece, in Cherokee County, Kansas.

(Photo courtesy of Minerals Unlimited)

Chapter 3   Crystals, Inclusions and Pseudomorphs

A crystal is the ideal mineral form, crystallization is the physical and chemical process that produced it, and crystallography is the study of both. Abbreviations are commonly used for these terms, "xl" for crystal, "xls" for crystals, and "xlized" for crystallized. The sides of a crystal are called "faces," the end is the "termination" and if the crystal has two ends separated by faces it is called "doubly terminated."

Most mineral substances, under ideal natural conditions, will produce a smooth multi-sided or faced solid form called a crystal. This is done through the process of crystallization as the mineral turns into a solid, from either a gaseous or liquid state. Crystals are produced, or may form, from hot gases, hot or cold solutions, or molten silicate melts under the influence of proper pressure and temperature. Crystallization is an interatomic force, and the resulting concrete external shape is directly related to the atomic structure of the mineral. Consequently, a crystal is the most important, definitive, and characterisic physical aspect of a mineral. Minerals seldom crystallize exactly alike, but any given mineral species adheres strictly to its own laws of crystallization irrespective of where it may be found.

Crystals occur in a fantastic variety of shapes and sizes, hundreds of them ranging from microscopic crystals to giants several feet in length. Although small crystals are the most desirable, practical, and inclined towards perfection, some giant crystal finds are of interest. For example, a single beryl crystal found at Albany, Oxford County, Maine, weighed 18 tons, was 4 feet in diameter and 18 feet long. In the Etta Mine, Keystone, Pennington County, South Dakota, gigantic single crystals of spodumene have been found, one 47 feet long, 5 feet in diameter and weighing close to 90 tons. A natural rhombohedron of calcite found near Helgustadir, Eskefiord, Iceland measured 20 by $6\frac{1}{2}$ feet, while the Iceberg Claim, Picuris District, Taos County, of New Mexico, yielded two calcite crystals 7x8x11 feet and 8x9x10 feet. Singles crystals of dozens of other minerals such as quartz, topaz, feldspar, galena, mica, gypsum, pyrite and barite have been found that weigh well over 100 to 150 pounds. Groups of associated crystals weighing several hundred pounds are not at all uncommon in many major mining ore bodies.

The casual observer of mineral displays in museums often thinks that natural crystallized mineral specimens have been cut and polished. True, fine specimens appear cut and polished, but this is purely a product of physical and chemical action.

An individual mineral crystal shape or form remains basically the same regardless of its size or where the specimen is found. For example, in Fig. 3-1 several quartz crystals are shown. They are of differing sizes and shapes and come from Arizona, Brazil, California, France, Italy, Japan, Nevada, North Carolina and Switzerland. Despite the apparent differences in size, shape and color, they all have exactly the same number of crystal faces—six on the hexagonal prism of the crystal and six on the termination—an hexagonal shape. And what is of much more importance, the angles between pairs of related faces are the same. Size, shape and color will vary, but the angles will remain the same. This applies to all minerals, not just quartz, provided crystals of similar habit or form are compared.

The study of crystallography is a very complex science by itself and is dependent upon mathematics, mineralogy, and chemistry. Crystallography deals with the varying physical solids, and the classification of the natural plane surfaces, called faces, seen on crystals. There are five subdivisions of crystallography—geometrical, structural, optical, chemical and physical. The average mineral collector has occasion to come in contact more with the geometrical, physical, and some optical aspects of crystals than the other divisions. The collector is primarily concerned with the study of crystal forms and appearances. The precise measurement of interfacial angles on crystals requires a laboratory piece of equipment called a goniometer, and few collectors pursue the study to this degree.

Due to the depth of the subject and its inherent complexity, crystallography is mentioned here only in very general outline, covering some terms and aspects most frequently encountered by the hobbyist.

As stated previously, a crystal is a natural, idealized, mineral form crystallizing in a highly distinctive and characteristic manner for each mineral. All but a few minerals have been found crystallized, or have been crystallized artificially in the laboratory. The shape of a crystal is determined by the molecules of which it is made. Each molecule of a substance has forces acting in definite directions, and molecules are attracted to each other in a pattern defined by these forces—a basic principle of atomic physics.

THE SIX CRYSTAL SYSTEMS AND SYMMETRY: Although an astonishing variety of crystal forms is known, all crystals fall within one of six "systems" according to the direction and magnitude of the internal lines of molecular force, or the position, number, and length of imaginary internal lines called "axes."

The six crystal systems are: (1) Cubic or Isometric; (2) Tetragonal; (3) Hexagonal; (4) Orthorhombic; (5) Monoclinic; and (6) Triclinic.

Within the basic six systems there are thirty-two special crystal "classes" which are distinguishable one from the other by various aspects of symmetry. In addition, many minerals within the six systems, and within one of the thirty-two classes, may show additional variation in the form of the crystal. Although the angles remain the same between the faces, the size of individual faces may vary, causing the mineral to assume a particular form or "habit." Some minerals have several "habits" of crystallization. Calcite, for example, crystallizes in the rhombohedral class of the hexagonal system, and at least twenty habits are known, more than any other mineral.

Although there is an apparently unending possibility of combinations between the basic crystal system, class, and form or habit, all basic differences rest with the "crystallographic axes," those factors which differentiate the six systems.

Crystallography defines the relative position of crystal faces, one to the other. Therefore straight lines are assumed to pass through the idealized center of any crystal form. Such lines are called the "crystallographic axes" forming the "axial cross" at the point of internal intersection.

The six systems differ through basic orientation of the axes and may be outlined as follows:

Isometric or cubic system: three identical axes intersecting at right angles.

Tetragonal system: three axes, two identical, perpendicular and horizontal to each other, and the third either longer or shorter than these and at right angles to them.

Hexagonal system: four axes, three of which are identical and intersect at angles of 60° in a horizontal plane; the fourth axis is perpendicular to the lateral plane of the other three and may be longer or shorter.

Orthorhombic system: three unequal axes intersecting at right angles.

Monoclinic system: three unequal axes, two intersecting at an oblique angle, and one perpendicular to these.

Triclinic system: three axes, all unequal and intersecting at three different angles.

As might be imagined, there is far more to crystallography than this simple outline of the six systems, which are not as simple as they may sound in some cases. It does, however, give some insight into the amazing orderliness of crystallized minerals, and at the same time, no attempt is made by the author to oversimplify the subject. No crystal is known which cannot be codified by the crystallographer as to its system, class, and habit or form.

Although the subject appears complicated, there is no reason why the hobbyist cannot master the recognition of the basic crystal systems, many classes, and the majority of common habit forms. If one wishes to pursue the study of crystallograhphy, standard textbooks on the subject should be consulted, as well as intensive personal examination and study of crystal specimens or crystal models.

The factor of symmetry is always present when observing minerals, and it is upon the study of symmetry that the six crystal systems were developed. There are two aspects of symmetry of interest to collectors; (1) plane of symmetry, and (2) axis of symmetry.

The *plane of symmetry*, or planes, refers to the theoretical division of a geometric solid into halves, with the resulting two sections identical in terms and number of faces, edges, and angles. Most geometric solids can be split more than one way to produce this effect. This is exactly like a mirror image. For example, a cube can be cut downward diagonally across the top, corner to corner, or through the middle of the cube. Place a half from either method of division before a mirror and the total appearance is a cube. A cube actually has nine planes of symmetry; *i.e.*, there are nine ways a cube can be cut to produce an identical mirror image. Most forms have many fewer planes of symmetry.

An *axis of symmetry* is interesting and fun to determine in well-developed single crystals. This refers to holding the crystal on a level and revolving it through 360° to see how many times the crystal repeats itself in appearance as it is turned. If during a revolution of 360° it appears to assume the same position twice, it is called *twofold symmetry,* repeating itself each 180°. If it assumes the same position three times during a revolution it is *threefold,* using 120° for each repeat, if every 90°, it is *fourfold,* and every 60°, *sixfold.*

It should be noted that the drawings accompanying textbook illustrations of crystals are idealized. They are perfect from the standpoint of equal faces, exactness of angles, and proportion. Absolutely perfect natural crystals with every face equal and matched are uncommon indeed. Consequently the term "perfect" crystal has two connotations. The first is its absolutely idealized proportions, the textbook and mathematical illustration; and the second is the "perfect crystal" of the collector which refers to a well-developed, complete, and undamaged crystal. The great majority of crystals found in nature involve some degree of distortion from being absolutely perfect. Distortion can be very minor, or to a point where the crystal is almost unrecognizable.

TWINS, PARALLEL GROUPING, AND DISTORTED CRYSTALS: Perhaps the most interesting of all crystal forms, and certainly among those most sought by advanced collectors, are specmens showing twinning or "twins." The subject of twinning can become quite involved from the standpoint of crystallography since each mineral has its own twinning laws. The recognition of many twinned forms is the product of study and experience on behalf of the collector. However, basically, there are two types of twins or twinned crystals.

In the first and most common type of twin, the whole growth of the crystal is such that one-half of it has developed in a reversed position from the other. Another way to explain such a twin is that one-half of it appears to have been revolved 180° around the other. This is called a "contact twin," as is well illustrated in Fig. 3-2.

The second type of twin also involves the 180° factor, but the two crystals are merged one in the other, or interpenetrate each other. Such a twin is called a "penetration twin," like the crystals in Fig. 3-3.

The foregoing refers to *two* crystals; however, specimens can show multiple or repeated twinning wherein there is an intergrowth of three, four, five or more individual crystals, and these are referred to as "threelings," "fourlings," and "fivelings."

Usually, crystallized minerals are found in groups of crystals of one or more minerals, and such groups are composed of individual crystals adjacent to one an-

Figs. 3-3, 3-4. At left is a fine penetration twin of transparent seagreen fluorite from the Heights Mine, Westgate, Weardale, Durham, England, now in the Robert Deidrick Collection. Specimen at right is not a twin, but a parallel grouping of many merged calcite crystals, from New Glencrieff Mine, Wanlockhead, Dumfrisshire, Scotland.

other. The overwhelming number of specimens appearing in collectors' cabinets or drawers are of this type and are simply called "crystal groups" or "crystal aggregates." Twin crystals, however, are sometimes easily confused by inexperienced collectors with a very common type of true multiple crystal grouping—that of the "parallel group." In this case two or more crystals of the same mineral have grown in such close association that they are actually intergrown and the axes of the crystals are parallel. In other words, this is a multiple group of crystals so closely merged that they appear as a somewhat complicated single crystal and often are mistaken for twins. However, the axes involved are all parallel, not inclined as is the case with all true twins. Note the specimen in Fig. 3-4. Quartz, wulfenite, galena, calcite, fluorite, barite, analcite, pyrite and copper frequently occur in closely-grouped, interesting and attractive parallel forms.

Two other mineral forms should be mentioned at this point since they, too, are sometimes confused with twins. These are rosettes and hopper-shaped or cavernous crystals.

The rosette form is peculiarly characteristic of the mineral hematite and is often referred to as the "iron rose" form. In such a case, small platy crystals of hematite group themselves in such a manner as to form a larger crystal and render a scaley, rose-like appearance to the specimen. Gypsum and barite are also commonly found as "roses." See Fig. 2-19.

Hopper-shaped or cavernous crystals are almost self explanatory. They contain major depressions in some or all of the faces and many minerals can show this form, particularly halite, pyromorphite, and vanadinite. Such forms are believed to be caused under conditions of fairly rapid crystallization, and often look like mere skeletons of crystals. Sometimes such forms are called "skeletal crystals." Specimens showing curved crystal faces are not uncommon, and a few minerals show this phenomenon frequently, especally diamond, quartz, sphalerite, dolomite, prochlorite and selenite.

The term "distorted" crystals speaks for itself. Almost all crystals found in nature show some degree of distortion or degree of imperfection from the standpoint of perfectly equated faces, and normal crystals can be subjected to a wide variety of crystallization conditions which can change, twist, or in some manner modify and distort the form. Crystals can be twisted, bent or seemingly squeezed, thereby creating unusual and interesting forms. Distorted crystals are common and some distorted forms are usually found within any major crystal deposit.

A famous locality that has produced some outstanding distorted crystals is in the lofty alpine peaks around Fellital, Canton Uri, Switzerland. Here, many smoky quartz crystals have been found that are strongly twisted. When viewed from the termination end they appear pitched in much the same manner as an airplane propeller. A group of radically distorted quartz crystals is shown in Fig. 3-5.

SCEPTER CRYSTALS: This is a fairly uncommon crystal form usually found in quartz, but several other minerals may show the characteristic form or shape. A scepter crystal terminates in a bulbous or larger termination than the body of the crystal itself. In other words, it resembles the shape of a king's scepter. In Fig. 3-6, two good examples are shown which well illustrate the extremes of scepterism: (1) the obvious expanded and almost doubly terminated scepter termination, and (2) a more-or-less pinching-in of the crystal to form the scepter.

INCLUSIONS AND PHANTOMS: One of the most fascinating aspects of the study of minerals and crystals is that of "inclusions." The term refers to one or more minerals "included" or occurring within another, usually a translucent or transparent host mineral. Inclusions do occur, in one sense, in minerals that are opaque, such as tetrahedrite in pyrite, sphalerite in galena, and dozens of others. These, however, are simply mineral mixtures or combinations extremely common in ore bodies, referred to as massive intergrowths.

To the mineral collector the term inclusion means a translucent or transparent mineral crystal in which other minerals occur as if suspended in space or frozen within the former. Inclusions (the included mineral) may occur as microscopic dust-like crystal particles, small or large single crystals, or more commonly as groups or aggregates of small well-formed crystals.

Figs. 3-5, 3-6. There are fifteen major crystals in the quartz group at left, all of them radically distorted. The specimen was found at Crystal Hill, Yuma County, Arizona. At right are excellent amethyst scepter quartz crystal from Guerrero, Mexico, and another, less radical type of scepter quartz crystal from Canton Grisons, Switzerland.

The most common mineral carrying inclusions is quartz. This is not surprising, since quartz is the most common transparent crystallized mineral found worldwide in ore bodies or mineral deposits. Inclusions may occur within almost all gem stones, as well as in minerals such as calcite, fluorite, barite, gypsum, halite and celestite. Any transparent or translucent mineral may occur with inclusions of another mineral within it. Inclusions are actually impurities, and they decrease the value of many gem stone minerals such as diamond, emerald, sapphire and zircon. However, inclusions may turn the commonplace quartz or calcite crystal into a collector's item of spectacular beauty.

Inclusions are not one of the best understood phenomena of mineralogy. Which came first, the inclusion or the host crystal? This is a perplexing question in some cases. For example, the gem mines of San Diego County have furnished some breathtaking specimens of clear quartz crystals completely penetrated by large raspberry colored crystals of rubellite tourmaline. Some of the penetrations are of such size as to cause wonder regarding which came first, the quartz or the tourmaline. Crystals grow by slow processes, and it appears that both the host and inclusion minerals grew simultaneously. During growth, crystals are primarily affected by (1) varying chemical composition of the depositing solutions or gases, (2) temperature, and (3) pressure. There are many variables involved, so that the entrance of another mineral into the major mineral crystallization is common and is probably a question of solubility of minerals in host solution. Examine a good collection of crystallized minerals and you will note that the majority of specimens comprise two or more associated minerals.

Wherever crystals are found there is a chance of finding inclusions, and in some particular localities crystals are almost always found with them. As mentioned, quartz leads the field in the number of included minerals. Quartz crystals have been found with striking inclusions of well crystallized pyrite (Fig. 3-7), rutile, hematite, garnet, tourmaline, cervantite, cookeite, chlorite, kaolinite, dolomite, stibnite, gold, amianth, azurite and malachite, to mention a few.

Quartz, calcite, fluorite and halite are all minerals in which movable bubbles of air or carbon dioxide have been found within a liquid inclusion inside a crystal. The most common liquid included is water, and if the cavity within the specimen is not completely filled, the remaining area creates a bubble. Bubbles may be minute or large and crystals have been found containing as much as a cup of water in a single cavity. In a few quartz crystal specimens from the state of Minas Geraes, Brazil, water bubble inclusions occur in which there are free suspended crystals of loose chlorite in the water itself. As the specimens are turned these crystals tumble about in their liquid prison.

In Brazil and Uruguay large agate nodules have been found that are in reality geodes filled with water, and give an audible splash when shaken. These nodular forms art called "enhydros." The geode localities around Keokuk, Iowa, have produced many enhydros, including a number filled with petroleum.

A related and interesting geode form are "rattle stones." It is not uncommon for an enhydros to 'dry out' leaving behind a loose crystallization or lump of mudlike material enclosed within the specimen. These geodes will rattle when shaken, hence the name rattle stone. Geodes that rattle have been found varying in size from that of a walnut to as large as a watermelon.

Calcite has also provided many superb inclusion specimens with notable examples of clear calcite enclosng native wire silver or aborescent and crystallized copper from the mines near Houghton, Michigan, and localities in Nevada. The Ojuela Mine, near Mapimi, Durango, Mexico, has produced calcite with some extraordinary inclusions of colorful minerals such as hematite, adamite, vanadinite, wulfenite, malachite and aurichalcite.

Inclusions and "phantoms" are discussed together since a phantom is a frequent type of inclusion. A phantom is a ghostlike inclusion which shows a former

Figs. 3-7, 3-8. The pyrite crystals at left occurred as inclusions in quartz, found at Belo Horizonte, Minas Geraes, Brazil. At right, a quartz crystal from the same source, now in the R. O. Deidrick Collection, shows series of over fifteen phantoms, one above another.

growth stage or period of development of the crystal. For example, during the growth of a quartz crystal a period may occur when kaolinite is introduced as an impurity in the solutions of crystallization. This admixes with the quartz leaving a pinkish coating on the outer faces of the crystal at that stage of development. For unexplained reasons such a coating will usually favor either the termination of the crystal or a few of the major faces. Sometimes the entire crystal is coated, but this is not too common. As the crystal continues to grow, the kaolinite impurities may disappear from the solutions and the crystal resumes normal clear growth, thereby encasing the kaolinite coating with clear quartz. This ghostly outline of the former crystal, as in Fig. 3-8, is a "phantom." During growth, impurities may be introduced a number of times and produce a series of phantoms, one atop the other. Brazil is notable for specimens of this type, and specimens have been found clearly showing two dozen distinct phantoms. Phantoms may appear only as a faint hazy outline, or may be thick and strongly colored.

Without a doubt some of the world's most spectacular and showy mineral specimens are of the inclusion and phantom type, especially those which show two or three different included minerals. Brazil, Switzerland and California have produced quartz crystal inclusion specimens of great beauty. Recent finds near Placerville, California, have been comprised of quartz crystals about the size of a pint cream bottle with brilliant bursts of acicular ruby red rutile emerging through a plumose base of green chlorite and extending well into the termination.

Inclusions and phantoms are interesting, not particularly rare in many cases, and are something the collector should be aware of when either collecting personally or purchasing specimens.

PSEUDOMORPHS: In Greek the term pseudomorph means "false form." In the mineral field there are many interesting false forms, or replacements, caused by the changing or alteration of minerals, yet allowing the specimen to retain its original crystal shape. In simple terms, a given mineral may have the crystal shape and form of another — a pseudomorph.

Pseudomorphs constitute one of the most fascinating, and sometimes misleading aspects of the mineral and crystal collecting hobby. Many go unrecognized by the beginner, while advanced collectors usually regard fine specimens as choice,

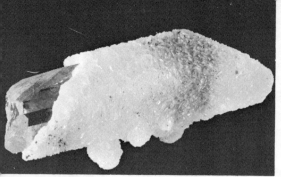

Figs. 3-9, 3-10. Some pseudomorphs begin with the process of encrustation. At left is a quartz crystal heavily encrusted by calcite and pyrite, from the Josephine Mine, El Dorado County, California. At right is an encrustation pseudomorph of pyrite after fluorite. The shell-like hollow cubic forms were fluorite crystals, later coated by pyrite and finally disappearing. Found at Nenthead, England.

due to the two-fold story each specimen represents. Not all of the unusual phenomena of pseudomorphic mineral forms is understood or has been explained, but there are four basic and different types as follows:

ENCRUSTATION PSEUDOMORPHS: These are the simplest of all pseudomorphs and are caused by the coating or encrusting of one mineral crystal by another mineral. In the beginning, one mineral coats another, such as the specimen in Fig. 3-9. in which a quartz crystal is coated with calcite crystals. Actually, these combination specimens are not true pseudomorphs, but represent a partial stage of potential development. Frequently, however, the original encrusted mineral crystal disappears, leaving behind a hollow pseudomorph shell, or a negative cast of the original mineral. Sometimes called "molds," such a form is a true "encrustation pseudomorph" and is well illustrated in Fig. 3-10. This specimen shows a series of cubic shells of pyrite where they originally coated fluorite cubes. The fluorite has disappeared, but it has left behind unmistakable evidence of its former presence.

ALTERATION PSEUDOMORPHS: The most common type of pseudomorph is the product of chemical alteration within a related mineral series, involving either the complete or partial addition of a new mineral or removal of original material. Here the specimen retains the crystal form of the original mineral, but by the process of alteration changes its chemical composition to another mineral. Such changes are confined to related minerals within an associated family group as among the many lead minerals, coppers, or iron minerals. A very common example is the replacement of pyrite by the related iron minerals goethite or limonite. A famous locality which has produced spectacular examples is Pelican Point, Tooele County, Utah. This locality has produced thousands of specimens, both as single crystals and groups, with some single crystal faces up to four inches across. A typical specimen appears in Fig. 3-11.

Other examples of alteration pseudomorphs are common among the following: anhydrite after gypsum, gypsum after glauberite, kaolinite after orthoclase, chlorite after garnet, and malachite after either cuprite or azurite. In Fig. 3-12, a classic example is shown, malachite after azurite. Malachite itself does not crystallize in the form illustrated and occurs in this specimen as an alteration replacement. Frequently such specimens show incomplete pseudomorphism, in other words, an azurite crystal shape; but perhaps half the specimen is still azurite and the remainder malachite.

REPLACEMENT OR SUBSTITUTION PSEUDOMORPHS: Occasionally a replacing mineral has no chemical relationship to the original mineral, but simply fills in a cavity, effecting a perfect cast of the former mineral form. Typical examples would be quartz after calcite from France, after anhydrite from New Jersey, and after fluorite from Cuba. The semi-precious gem stone mineral crocidolite or tiger-eye from Griqualand West, South Africa, is actually a pseudomorph, and is a quartz replacement after crocidolite asbestos.

PARAMORPHS: These are uncommon and the most difficult to understand of the pseudomorphic forms. Paramorphs occur only within polymorphous minerals—those few that are chemically identical but crystallize into distinctly different forms such as calcite, rutile and brookite. In a paramorph the chemical composition remains unchanged, but the specific arrangement of the atoms has been modified. The most well known examples are the rutile paramorphs after brookite from Magnet Cove, Arkansas.

"SKULLS": The term "skull" refers to a distinctive form closely related to an encrustation pseudomorph, and characteristic of certain native copper specimens from some localities, notably the Keweenaw Peninsula of Michigan. A very good example of a hand-size native copper "skull" appears in Fig. 3-13. These interesting forms are the result of native copper forming an arborescent sheet around a pebble in a conglomerate rock formation of a copper ore body. These skull or cup-like specimens vary from about the size of a fifty cent piece to some as large as buckets. Thin walled native copper skulls will ring like a cow bell when struck.

FOSSILS: Fossils are the replaced mineralized forms of former plant or animal life. All fossils, except those preserved in either amber, tar, or similar medium, are actually true and complete replacement pseudomorphs. The most common minerals involved are silica, or quartz in its various amorphous forms such as chalcedony, jasper, opal and calcite. In many unique localities the replacing agent is some other mineral, and such specimens are usually of interest to the collector.

In Fig. 3-14, two typical examples are shown. The larger specimen is a small tree limb section from the Copper Queen Mine, Bisbee, Cochise County, Arizona, and is completely replaced by native copper. The snail shell is from a locality near Galesburg, Illinois, and is of solid marcasite. Around 1908 the Kelly Mine in the Magdalena Mining District, Socorro County, New Mexico, was the source of fossil mollusk shells and crinoid stems completely replaced with magnificent apple-green smithsonite.

Figs. 3-11, 3-12. At left is a pseudomorph showing complete replacement by alteration process of limonite after pyrite; found at a locality famous for such specimens, Pelican Point, Tooele County, Utah. The specimen at right is malachite replacing azurite crystals, i.e., malachite pseudomorphs after azurite; from Cochise County, Arizona.

Chapter 4

Where To Collect Specimens

The odds are all against a novice merely selecting a likely looking mountain and starting out to fill a knapsack with crystals. It just isn't that easy. However, knowing a little about the environment and conditions under which crystallized minerals form, as well as logical places to look for specimens, will certainly eliminate a great deal of territory.

There are millions of tons of rock exposed in the vast expanse of mountains and deserts of the planet Earth, yet only an infinitely small proportion or number of crystals exist. Despite such a seemingly overwhelming disproportion, it is surprising how readily good specimens may be found once the collector becomes familiar with the fundamentals of mineral and crystal hunting. Like anything worthwhile, a little study, patience and experience will pay dividends.

There are very few areas in the United States, or anywhere else for that matter, where crystals of some sort cannot be found. In each area, the abundance varies and the methods of hunting and obtaining specimens may differ considerably. In some localities crystals can be had merely by picking them up from the ground where they have eroded from their rock matrix. In other localities, specimens may be obtained only after hours of hard work in rock matrix such as basalt, pegmatite, or quartz outcropping.

Many outstanding crystal specimens have been found by amateur private collectors. The chance of finding quality specimens is always present. Efforts of many local mineralogical and gem societies and private collectors have greatly contributed to the wide variety of crystallized specimens which have been found in the United States.

CRYSTAL ENVIRONMENT: Geologically, where do minerals crystallize? First of all, crystals can occur in all three basic categories of rock; igneous (rock of molten or magmatic origin), sedimentary (rock composed of solidified sediments), and metamorphic (rock that was formerly igneous or sedimentary but has been altered or changed by heat and/or pressure). As outlined in the previous chapter, minerals crystallize under ideal conditions from molten silicate melts, gaseous vapors, and hot or cold super-saturated solutions. The latter is the most important source of crystallization. Many minerals crystallize within a host rock formation and the crystals are found imbedded as an integral part of the rock matrix itself. Feldspar or garnet crystals, for example, are commonly found as component units making up an overall rock mass. They are usually harder than the surrounding rock, or less subject to

Figs. 3-13, 3-14. Specimen at left is a "skull" of native copper originally formed around a pebble. It was found at Houghton, Keweenaw Peninsula, Michigan. The two unusual fossils at right are a small tree limb completely replaced by native copper, from Bisbee, Arizona; and a shell replaced by marcasite, from near Galesburg, Illinois.

weathering, consequently they may erode out of bedrock as loose crystals. Most crystallized minerals are deposited in some sort of opening or cavity in rock. They may be deposited as a "primary" material, which means they formed at the same time as the surrounding rock, or as a "secondary" deposit, meaning they crystallized after the host rock or matrix was created.

The openings or cavities in rock comprising the crust of the earth are formed in various ways. Cavities which are called "vesicular" in origin are formed by bubbles in molten volcanic rock. Such bubble cavities may be very tiny or quite large. Frequently rock masses show the effects of shrinking as the original rock contracted during cooling. This may cause openings called "miarolitic" cavities, and these, too, may be large or small. Shrinking can also produce large cracks, fissures or joints in rock masses over a large area.

Similar to the cracks and fissures produced by shrinking are those produced by diastrophism, or faulting. The earth's surface is comparatively quiet nowadays, except for a few volcanic vents and earthquakes. But, the crust is well fractured and faulted as through the millions of years it has gone through countless upheavals and major shiftings of land masses and oceans. Faulting can produce a cavity in the earth by moving both sides of the fault apart. Another fault-produced crystal location is called "fault gouge." This is a faulted area where the earth movements have created a highly fractured zone composed of a jumbled mass of rock fragments. Space exists between the angular fragments of "gouge" and secondary crystallization may occur there.

Crystal-lined cavities in rock and ore bodies are usually referred to as "pockets" if they are large, and "vugs" if they are small, although the terms are used interchangeably.

The remaining type of cavity is produced by secondary action itself. Ground waters, either hot or cold solutions percolating upward or downward through rock bodies, may dissolve out cavities, tubes, smooth fissure-like areas and caves. Limestone caverns are the product of this dissolving type action. The limestone rock is dissolved by waters thus creating the cavern areas. Next mineral laden waters seep into the cave to evaporate and leave their mineral deposit behind as a stalactite, stalagmite, or other crystalline or crystal form. Calcite, aragonite and gypsum are often found well-crystallized in caverns.

All of the cavities, cracks, fissures or cavernous openings mentioned are potential environments for crystal deposition. Gases or super-saturated solutions under tremendous pressure may flow or be forced into these openings, there to crystallize. Normally, crystals adhere to the sides of openings, or in the case of gouge, coat the material. It is also common for an area such as an ore body to experience several successive stages of mineralization, or for the process of crystallization at any one time to involve more than one mineral. Several related minerals may crystallize in an ore body, or if there have been successive stages of mineralization, unrelated minerals may be found associated with each other. Crystallized minerals are valuable clues to the history of an ore body or outcrop and consequently are often valuable to mining geologists in determining the sequence and possible extent of mineralization in a mine or mining district.

COLLECTING SITES OR SOURCES: There is no precise method of determining exactly where crystals may be found, and this adds to the excitement of collecting. There are several basic sites or sources where the collector may logically search for specimens.

In the following pages several of these sites and sources are outlined. Keep in mind that some promising-looking localities may prove barren and others, not so attractive, may be outstanding. The exterior appearance or setting of a locality has nothing to do with the possibility of specimen material. Also, specimens may be found where you least expect them. For example, some unusual localities in the

State of California would be: agate geodes in a foundation excavation on Spruce Street in Berkeley; chiastolite crystals from behind the Beverly Hills Hotel; tourmaline and quartz crystals from a corral in Hemet; fine barites in a home subdivision in Palos Verdes; gold nuggets from streets in Placerville and Sonora; twinned gypsum crystals from a road cut in Sunland; and garnets from the diggings of fence post holes at Tiburon.

Like the old saying about gold, crystals are "where you find them." The foregoing represents some odd and unexpected localities for specimens, while those following are more logical or common. Remember, whenever possible, permission should be obtained to visit and collect on property marked as private.

Although this might sound like the proverbial circle, collectors often obtain the best information or leads to localities by asking other collectors. Most collectors are willing to share locality data with others of the same interest. Corresponding with collectors belonging to societies in other states will usually be a very real help if a trip into their area is anticipated. Another lead for localities is found on the labels accompanying specimens in many private and public displays. Many a collector has located an isolated area that proved to be a bonanza by beginning with the locality data on a label.

In general, those localities within a mining district or heavily mineralized area will prove the most interesting and rewarding. The West is populated with dozens of old ghost towns. Here the collector may search the dumps, test pits, tramway areas and prospectors' holes. Such "ghost towns" are always subject to reopening. Even though its buildings may be old and weathered and the machinery rusted, remember they still belong to someone and collecting there is a privilege.

MINES: The majority of fine crystallized specimens seen in museums and the better private collections have been found in fissures, pockets, or vugs in mines. It is surprising that in most instances these crystals are of little or no commercial importance to mine operators and are destined for the mine dump. Exceptions would of course be gem stones. Miners removing commercially valueless rock from mines have brought to the surface the world's finest mineral specimens. Miners hired by dealers to remove mineral specimens have made hundreds more available to the private collector.

Collecting in a mine proper, with the exception of open-pit mines, is seldom done by the private collector unless special permission has been obtained to enter a mine for this purpose. Exploring and collecting in old abandoned mines is quite dangerous. Mine timbers shoring up shafts, stopes and tunnels will decay if not maintained; ladders will become rotten and unsteady (with perhaps a few rungs missing), and the air supply in non-operating mines may be foul or even dangerous from the accumulation of toxic gases Unless the collector is very familiar with mines and mine safety, it is best to leave the mine collecting to someone else! Abandoned mines are favorite haunts of the black widow spider, scorpion and rattlesnake, especially near the entrance. There are occasions, however, when a collector may have a chance to be guided into a mine with collecting privileges, and this can be an unforgettable experience.

MINE DUMPS: Recognizing that mine dumps are the next best thing to the mine itself, mine dumps or waste piles throughout the nation have been favorite hunting and collecting sites for many years.

Not all mine dumps are the same in terms of their purpose or the type of material they contain. There are three different types of dumps commonly found in mining districts. They are as follows:

(1) *Mine dump,* or *waste dump:* This is composed of unwanted rock materials considered valueless to the operation. Such a dump, as appears in Fig. 4-1, is composed of unsorted sizes of material ranging from fragments to angular blocks. This is the most common type of dump and the one most likely to yield specimens.

Fig. 4-1. Head-frame and waste-dump of the Kennedy Mine in California, once deepest in North America. (Calif. Div. of Mines photo)

(2) *Stock pile*: A stock pile, which looks like any other dump, is composed of usable ore that is piled waiting shipment to a precessng plant or mill. These may be interesting sources of material, but specimens should not be removed from an operating mine stock pile without permission or offer of payment for the specimens. Stock piles mean cash to mine owners.

(3) *Mill dumps*: These dumps or "tailing piles," are composed of sorted and processed material ranging from powder to heavy gravel-like crushed rock material. This material, as in Fig. 4-2, has all been through some sort of mill processing, and is not likely to produce specimens. Mill dumps are distinctive because of the evenness of the materials on them. Mills were moved from one place to another, so a mill dump may exist where there is today no evidence of a mill building or equipment.

Fig. 4-2. Mill dump "tailing piles" from the famous Comstock Lode, Virginia City, Nevada, showing material which has been processed through mill.
(Calif. Div. of Mines photo from Combination Shaft)

Why are mine dumps good places to hunt? The first point is obvious — the dump is composed of rock materials brought from far underground, making available to the collector a source he cannot reach by other methods. Whatever areas of mineralization the mine has penetrated will be represented by materials on the mine dump. Generally speaking (there are many exceptions) the mine sections nearest the surface may contain various oxidized, alteration, or carbonate minerals which are often well crystallized. As mines penetrate this upper zone they reach sulphide layers which are richer in ore but usually contain fewer crystallized specimens. Thus, mine dumps containing material excavated during the initial operation of the mine are particularly good collecting spots, while the later commercial workings are below the richest crystal producing area. Such collectng sites may be exhausted of surface specimens quickly.

Digging into a mine dump is a good technique to keep in mind. Specimens on the surface of a dump may be bruised and damaged but digging may reveal fresher material, or increase the chances of finding undamaged specimens. A dump is actually composed of different layers piled one atop the other. Specimens may not appear on the surface, but dumps appearing totally barren of specimens have yielded many fine specimens to the collector willing to do a little shovel and pick work.

Sometimes good specimens reach the mine dumps simply because they were missed while going through various screening, grading or sorting processes. For this reason many very fine specimens of tourmaline, beryl, topaz and kunzite have been found on the many gem-mine dumps of San Diego County, California. The crystals may have been coated with another mineral that has oxidized or eroded away while the specimen lay on the dump. In a few localities, notably the Himalaya Mine, Mesa Grande, California, the dumps in the past have proved so lucrative that collectors searching for tourmaline have actually put them through screens. It is doubtful that there remains a square foot of the Himalaya dump that has not been screened at least once. Screening as a collecting technique is often overlooked by collectors, yet can be used successfully under a variety of conditions.

Usually, better specimens may be found on dumps, either large or small, if they are remote from main highways, since these are less likely to have been picked clean by other collectors. After all, collecting minerals has been an important hobby for many years and collectors generally like to travel.

Experienced collectors may also "work a dump" with a sledge hammer breaking up large boulders. Small and fresh pockets or vein sections are often encountered in this manner. Be careful not to start a rockslide.

It should not be assumed that all mines or mine dumps contain specimens, for such is not the case. Many mines contact little or no crystallized material; still, mine dumps remain one of the most logical and likely collecting sources, especially in the West where there are countless hundreds of them.

TRAMWAYS AND LOADING PLATFORMS: The tunnel or shaft entrance to a mine may be some distance from the main mine buildings or the area where the ore is either processed or shipped. Ore may be brought into the depot by aerial bucket tramway, ore car trains or by truck. Consequently, another possible source of specimens is around the tramways and the related loading sheds and platforms. Material may spill from ore buckets on tramways, especially where they may be bumped or swayed at a tower or swung over the crest of a ridge. Material may spill from ore cars and trucks around loading platforms or sheds. Loading areas in the Castle Dome Mining District, Yuma County, Arizona, have produced interesting fluorite crystal groups, and the tramway of the Gold King Mine, La Plata County, Colorado, has yielded a variety of specimens. If you want to find out what type of mineral was primarily mined at a mine, look around the loading platforms first, not the dump. Remember, the dump contains material considered worthless to the operation.

Fig. 4-3. Typical open-pit copper mine, owned by the Phelps-Dodge Company at Ajo, Arizona, which has been worked since the 1850's.

Even the road leading to a mine may be a possible collecting source if the ore was trucked from the mine to some other location. Again, material may spill off a truck, particularly on rough and bumpy roads. Specimens found by such means are few and far between to be sure, but some excellent specimens have been found that obviously fell off a truck during ore shipment.

OPEN-PIT MINES: Open-pit, or "glory-hole" mines are another excellent source of specimens. Permission is almost always necessary for entry, but many operators have been very considerate and cooperative in allowing collecting at times and in sections where collectors will not interfere with the operation of the mine.

Open-pit mines are usually operated within large low grade ore bodies for either iron or copper ores. Open pits are potentially good localities since their bench-like walls expose miles of mineral bearing rock. In the West many of the huge open-pit copper mines in Arizona, such as in Fig. 4-3, Utah, Nevada and New Mexico have furnished an abundance and variety of fine specimens. Likewise, many outstanding iron, copper, and silver-bearing minerals, as well as fine specimens of calcite have come from the open-pit and strip mines of the Keweenaw Peninsula, Michigan.

Fig. 4-4. Large "cut" on hill at end of this road is the huge "glory hole" of the Morgan Mine at Carson Hill, Calaveras County, California. A solid gold mass worth $73,000 was once taken from this mine.

As is the case in quarries, collecting may be done in rubble on the floor of the pit or along its benches as well as in the exposed walls of the benches.

In various mining districts in California (Fig 4-4), Arizona, Idaho, Colorado, Nevada and New Mexico, huge deep pits called "glory holes" have been sunk in search of lead, silver, gold and copper. Such mines operate within a localized and highly concentrated ore body close to the surface of the ground. Glory holes are frequently so precipitous that collecting on the walls is hazardous.

PLACER MINES AND DREDGE TAILINGS: Placer and dredge areas are not particularly lucrative hunting sites, however, occasionally interesting specimens are found. Placer or hydraulic mining was formerly a method of major importance in some western states. Today the big placers are gone and the havoc they brought to the foothills and countryside is being slowly softened, as some trees and undergrowth take root in the great hillside cuts and scars blasted by the powerful water jets of the hydraulic pumps called Monitors. A typical placer area is shown in Fig. 4-5 and has been the source of loose, water-worn crystals of pyrite, limonite, garnet, epidote and quartz.

Similar to placers in terms of the type of area mined are the "tailings" or waste piles left to the sides of a gold dredge as it ate its way up a channel, river or man-made pond or lake. In Fig. 4-6, an aerial view is shown of a typical tailings area depicting miles of caterpillar-like tailings of the gold dredge.

QUARRIES: Rock quarries (Fig. 4-7) and some types of gravel pits are potentially good hunting spots for mineral collectors. The geologically complex formations of many limestone, silica and rock quarries have provided hundreds of outstanding specimens, particularly in the eastern states. A magnificent array of various crystallized minerals have come from the different trap-rock quarries of New Jersey and New York State. California's Crestmore Quarry at Riverside has revealed over one hundred and ten different minerals.

In a quarry, two sources may be explored for specimens: the rock rubble on the floor of the quarry and the quarry face itself. Pockets, veins or seams of minerals are often encountered. The best collecting in quarries is usually facilitated by heavy tools, particularly sledge hammers and large cold chisels.

Many quarry operators have courteously extended collecting privileges to

Fig. 4-5. Entire area of these placer diggings on San Juan Ridge, Nevada County, California, was blasted out by hydraulic Monitors.

Fig. 4-6. Thousands of square feet of dredge tailings are seen in this aerial view over an area of western Mariposa County, California.

Fig. 4-7. Limestone quarry of Diamond Springs Lime Company, El Dorado County, California. There are hundreds of similar excavations which are possible collecting sites.

(Calif. Division of Mines photo)

hobbyists and allow collecting in non-working sections of the quarry or in the quarry itself on weekends, after hours, or holidays. Some operators request visitors to sign a release form before entering the quarry, absolving the owners of responsiiblity in case of accident.

Abandoned quarries should be approached with caution, especially if they are water filled. In either abandoned or operating quarries the collector should always be alert to the danger of falling rock.

Gravel pits usually operate in a loose float material or overburden, although some operations may be in solid rock that is extracted and crushed for gravel. Pits may contain some float specimens as water worn pebbles, agates, fossils and massive mineral forms. Rarely, a gravel pit may have a secondary deposit of minerals which may have crystallized in veins or around pebbles and boulders. Pits that operate in hard rock constitute the same collecting conditions found in other quarries.

OUTCROPS: Rock outcrops are another logical place to hunt specimens. Outcrops are protrusions of rock which have either been pushed upward through the local bedrock, or being of relative hardness, have been left as a remnant of erosion. They may occur in any terrain, the open plain, desert, mountain side or heavily forested hill. Outcrops may be small, perhaps only a few square yards in area, or they may extend for several miles. Usually they stand out in contrast with the surrounding bed rock and terrain by virtue of a different color, an obvious protruding rock mass, or a thin sharp ridge.

An outstanding example of an outcrop appears in Fig. 4-8. This is a thick white quartz outcrop, bearing silver and gold, comprising a section of the Great Silver Belt, Tioga Hill Mining District, California. It extends for about one mile. In such an outcrop, specimens might be encountered within the quartz mass itself as well as along the sides or "contact zones" where the outcrop rests against the bedrock. The same collecting possibilities would apply to the limestone outcrop showing in Fig. 4-9.

Another locality consisting of outcrops is around the head of the San Benito

Figs. 4-8, 4-9. At left is an excellent example of quartz outcropping on Tioga Hill, Yosemite National Park, showing old incline shaft and buildings in background. This "Dana Village" of the 1870's is now within the park where collecting is prohibited (D. Hubbard photo). Limestone outcropping near Springfield, California, seen at right.

Fig. 4-10. Pegmatite dikes lace the mountainsides near Rincon, San Diego County, Calif., in an area productive of fine gem and specimen material of tourmaline, beryl, topaz, kunzite, garnet and quartz. (Photo by Mary Hill)

River, San Benito County, California. Here the low chaparral-covered serpentine hills are broken by outcroppings of hard schist in which a wide variety of uncommon crystallized minerals are found.

DIKES AND VEINS: Closely related to outcrops are *dikes* and *veins*. Dikes usually look like what they are — filled-in cracks in the bedrock. Dikes are of molten magma origin, and are extruded (forced) into older rocks following joints, fissures and cracks. They differ in composition from the host rock and can be important sources of minerals. The gem stone mines of Maine and San Diego County, California are located in a coarse dike material called pegmatite. Crystallized minerals are found within pockets inside the dikes as well as along the contact zones.

Dikes that are the source of fine mineral specimens are found in many localities in the Rocky Mountains of Colorado, especially around Pikes Peak, in many areas of the Black Hills of South Dakota, and numerous localities in Brazil, Russia, Africa and Madagascar.

The term *vein* needs little explanation. It refers to a concentration of mineral ore in which pockets are often located. Dikes and veins can be small or large, some dikes reaching a thickness of eight or ten feet. In the immediate mining area around Pala, San Diego County, there are some four hundred easily-recognizable pegmatite dikes which stand out like cream colored ribbons on the brown hillsides. See Fig. 4-10.

Dikes or veins may also fluctuate in their dimensions. For example, in the Borrego Badlands of southern California thin veins or veinlets of calcite (sometimes called "stringers") can be followed through rough country for hundreds of yards. Occasionally these tiny veins pocket into huge vugs lined with crystals. This locality was extensively prospected and mined during World War II when optical grade calcite was a high priority item. The method used to explore this area is one any collector can use — merely following the veins. This technique is applicable to many badland or desert country localities as well as mining districts in several states. It does not take a large vein or dike to yield good specimens.

CLIFFS: A large exposure of rock surface such as a cliff is a possible hunting spot. Cliffs are logical sites whether at the seashore, in the desert, high in the mountains or bordering a stream or river. The fallen rubble of rock at the base of a cliff or on a mountain side is called *talus*, and searching through this gives a good clue or cross section of what is above in the cliff face or on the mountain side.

Figs. 4-11, 4-12. At left, flats of desert areas are littered with "float" which gives a sampling of minerals in hills and peaks beyond. At right, the three circular specimens are well-cores from Searles Lake, California. From the M. Vonsen Collection, they contain thenardite and trona crystals. (Photo courtesy California Academy of Sciences)

FLOAT: Similar to talus, but somewhat more removed from the original source is *float*. Float is the rock material carried in canyons and down hillsides by water Tons of float appear in Fig. 4-11. Whatever is found in the float came from somewhere upstream. In this illustration, the rock material in the foreground came from the cliff faces and high mountain sides in the distance. Many fine localities have been discovered by collectors who carefully followed the float up canyons or stream beds until they found the original outcrop or source of the material.

Usually float materials are waterworn and eroded, and the degree of abrasion (how much they are rounded and worn) gives some indication of the distance they have come. Not all float can be traced to a local source, since it may be of glacial or flood origin and the specimen might have been carried for many miles in centuries past. There is also the possibility that a float specimen has originated from a very small deposit and your lucky find may be impossible to locate in a large area of difficult terrain.

WELL CORES: An unusual and limited source of crystallized minerals is within well cores, the round tubular cores of rock material obtained from the center of core-type drills used in some types of mining. These are primarily employed in very soft rock or hardened mud and silt materials, such as found in many dry lakes of California, Texas and Nevada. Such mines usually operate for sulfur, borates and sulfates. In Fig. 4-12 three "cores" from wells at the famous Searles Lake, San Bernardino County, California, are shown. These are sodium minerals, trona and thenardite, and show clearly as crystallized pockets or layers through which the probing well drill passed.

SILICIFIED LOGS: Strangely enough, silicified or petrified logs may be the source of good crystallized mineral specimens. Fossil logs, which from the mineralogical standpoint are pseudomorphs, may contain pockets or cavities lined with crystals. This seems to be particularly true of logs that are poorly silicified, or logs that are not solidly replaced with hard chalcedony, opal or jasper. Many well known fossil wood areas are of the poorly silicified type and consequently of little interest to the gem hunter and lapidary. Also, silicified logs of low quality may occur within fossil forests of fine quality.

In Fig. 4-13, a typical fossil forest of this type appears. This locality is in the

Figs. 4-13, 4-14. At left, strata in Tom Miner Basin, Montana, includes huge upright stump in foreground as well as many crystal-lined fossil logs and caves. At right, road cuts intersect different geological formations. Road cuts may reveal specimen material, clues to collecting sites.

remote Tom Miner Basin of the Gallatin Mountains, southwestern Montana. The strata is volcanic breccia which buried a series of forests, one atop the other. Many of the logs contain cavities in which crystallized fluorite, calcite, celestite, barite, dolomite, aragonite and quartz have been found. A similar region in the badlands north of Cameron, Arizona, has produced fine plates and groups of small, brilliant, smoky quartz crystals, and clear quartz with hematite inclusions.

CUTS AND TUNNELS: Highway, road, railroad and dam cuts, as well as tunnel and foundation excavations comprise another group of sources not to be overlooked.

Like mines, cuts expose rock material below the surface of the ground and specimens have been found here. For example, a back-country mining road in the La Plata Mountains, Colorado, exposed a small but fine deposit of garnets; a railroad cut near Afton, California, revealed good stilbite crystals; a fire-road on Mt. Tallac in the Sierra Nevada was almost paved with pyrite crystals at one point; and a few extremely fine pink tourmalines were turned up by the blade of a road grader as it popped open a pocket of crystals in a roadside dike near Rincon, California.

Collecting in road or railroad cuts must always be done with due consideration for defacing the cut and causing material to fall onto the road or highway. If a road cut has exposed interesting material, there is a strong possibility that similar material may be nearby on hillsides where collecting would do no harm. See Fig. 4-14.

In a few isolated cases, foundation excavations for buildings have produced good specimens. A notable find was within the confines of New York City in the excavation for an apartment house at 96th Street and 4th Avenue many years ago. The foundation excavation revealed pockets of excellent, semi-transparent, white microcline crystals, some as large as two inches across.

The rock removed during the excavation for a tunnel may prove interesting. Tunnel refuse is dumped just as a regular mine operation would dump, and collecting conditions are the same. Tunnel dumps cannot be considered a major source of specimens, but should never be overlooked.

Three tunnel excavations in particular have produced well-known specimens. The Simplon Tunnel under the Swiss Alps encountered many well-crystallized pockets, notably complex curved groupings of dolomite. The Black Rock Tunnel in Pennsylvania intersected many veins and seams bearing crystallized material of which the ankerite on quartz specimens were among the most outstanding. A tunnel for the Los Angeles Water Aqueduct near Banning unearthed hundreds of pounds of handsome garnet specimens.

In discussing areas for collecting, two very important matters should be mentioned. The first is safety, and the second, permission to collect.

SAFETY: Needless to say, mines, quarries and mountain sides are potentially dangerous locations and caution should be exercized at all times in the field. Don't roll boulders down the dump, quarry face or mountain side without first seeing who is below. Keep a sharp eye out for rattlesnakes and black widow spiders, and be careful with fire. Also watch out for rusty nails around old mine buildings. Go prepared with adequate equipment to collect specimens. Sometimes the simplest accessories, particularly an eye shield and heavy gloves, will prevent many minor and uncomfortable accidents while collecting.

Don't take a wild swing with a sledge hammer or pick without first looking to see where your companions are. Not only is there the danger of striking someone accidentally with the tool, but chips of rock from a strong sledge hammer blow can inflict a nasty cut.

If you are visiting a remote collecting area, always let someone know where you are going. Travel with companions, never alone, and take adequate gasoline, water and emergency food supplies "just in case." A good shovel and an axe can be as important as a carburetor and a battery, and should be considered standard equipment in the car. Carry a flashlight with a spare set of batteries.

Don't wander around an old mining district at night. You may find some shaft you didn't expect to! Mine heads or shaft openings at the ground level may be overgrown with underbrush or covered with planking that is now rotten.

If you have been given permission to collect in an active area such as a quarry, be sure to stay out of the way of the workmen, machinery, trucks, tramways and conveyors. Collect only where you have been told it is safe to do so, and leave anything alone you might think is a stick of dynamite, dynamite cap or fuse.

Carrying your own canteen of water is better than drinking from streams and pools in many areas. Of course, in mountain country a fresh, rushing, cool stream is quite safe (unless full of ashes from a forest fire) and is probably better water than you brought with you. However, water around mines, particularly if running out of mine diggings themselves, will proabbly be highly mineralized and may be dangerous.

Appropriate clothing, especially foot and head wear, is important. Some sort of hat should be worn if collecting is done on a hot day. Some people prefer canvas shoes in the field since they are comfortable and give good foothold in rocky areas. However, such footwear is easily penetrated by sharp rock fragments, sticks and cactus. Low boots or heavy work shoes are more practical. Carrying a scarf or bandana is also recommended.

A first-aid kit should be in the car. In addition to the normal items, it should contain a snake bite kit, salt tablets, eyewash and eye cup.

PERMISSION TO COLLECT: There are hundreds of localities where no permission is needed to collect, but there are just as many where it is required and should be obtained. In recent years, the subject of respecting private property, even if on a mountain top or in the middle of a desert, has become a point of increasing concern to all outdoor hobbyists and sportsmen.

Under no circumstance should specimens be collected in a National Park, National Monument, or any other area administered by the National Park Service.

It is against federal regulations and the Antiquities Act to collect in these areas, and these regulations have been enforced in several cases. Service areas have been set aside for the pleasure and enjoyment of generations to come, and the National Park Service needs the understanding backing of all citizens, particularly those with scientific and conservation interests. It is also against the law to collect in State and local public parks.

If collecting is to be done on private property or in posted areas, it is always a good idea to obtain permission. Most property owners will appreciate the courtesy of being asked, and reciprocate with permission and may provide valuable information on the best spots for hunting specimens. However, it is unfortunate that many good collecting areas in the United States have been closed to collectors by property owners because of thoughtless acts on the part of both game and mineral hunters.

Many localities, especially in the West, are in cattle or sheep grazing country where gates left open, broken fences, untended campfires and littered campsites may spell financial loss to the owner and discourage further collecting on the land. A few mines formerly accessible to hobbyists have been closed to collectors as the result of vandalism in mine machinery and buildings, boulders thrown down mine shafts, plus outright theft. *In such ways, a few thoughtless individuals have spoiled collecting opportunities for all.*

Local inquiry, particularly of collectors or the local mineralogical club or society, will provide facts regarding collecting restrictions on a given locality. Sometimes a nominal charge is made by a landowner for access to a privately owned area.

A factor that has antagonized many mineral collectors, societies and property owners is the "specimen hog." Localities have been picked and blasted clean of specimens by the so-called "collector" who takes them away by the truckload. One notable area was closed by the owner after two of his cattle had to be killed as a result of falling into collectors' blast holes.

Take only enough specimens for yourself, perhaps a few for fellow collectors and trading, but leave something for the next collector to find and enjoy. Likewise, if you locate a good specimen but lack the proper tools to get it out, it is better to leave it for someone else.

Some lands have been legally claimed by mineralogical societies as a legitimate move to prevent the locality from being looted for commercial purposes. Such society claims have restrictions on them such as permission to collect, a limiting of the number of pounds of material that may be removed, and a closing of the locality to all commercial collectors.

Some individuals have also legally claimed collecting sites and operate on a similar basis. In one or two cases however, individuals have carried this activity to a ridiculous extreme, claiming every rock outcrop for hundreds of yards where any sort of specimen might be found. Unscrupulous individuals have also moved in and "claimed" a few localities where hobbyists have collected for many years.

Chapter 5

How To Collect and Prepare Specimens

In order to collect minerals and crystals efficiently, a few somewhat specialized tools are essential in addition to the standard geology pick or hammer. Without them it may be impossible to obtain a specimen. Attempting to remove a large calcite or quartz crystal group with a household hammer, jack handle, or another rock seldom gives good results.

One of the most common failings of collectors is to go into the field with inadequate equipment, especially tools to do work heavier than anticipated. A very few collectors may employ the use of dynamite — but needless to say, this is definitely for those who are familiar with explosives and their correct use.

With the exception of the geology pick or hammer, related collecting equipment can often be obtained at war surplus or used tool stores at considerable savings. Used tools will do just as good a job as new ones.

In Fig. 5-1 a variety of useful tools are shown. The geology picks were purchased new, the remainder at secondhand tool outlets; the total cost for all equipment shown was under $3.00. Appearing in the illustration are:

1. Two geology picks or hammers, one with a pointed end, the other with a chisel end.
2. A small five pound sledge hammer.
3. A $2\frac{1}{2}$ foot chisel pointed steel bar called a "moil." These come in longer lengths also.
4. A carpenter's bar.
5. A short, broad faced shovel.
6. Ditch digging pick with one side a bladed edge.
7. An assortment of different sized cold chisels, and dental tools.
8. One pair of nippers.
9. A garden claw-type hand cultivator.
10. A "pocket scraper."
11. Sifting screen, ¼" mesh.
12. Eye shield, or eye glasses.
13. Heavy gloves.
14. Small hand lens.
15. Pocket compass.

Of course, not all these tools would necessarily be carried to the collecting site, but they encompass the span of useful tools and accessories that should be taken with you in the car. If you find you need heavier or longer tools, you have them with you. Also, arriving near a locality, you usually are able to anticipate your possible tool needs. In such cases the shovel and large pick are more valuable in the trunk of the car than back home in the garage or tool shed.

The use of most of these tools is obvious. The most important item is the mineral or geology pick or hammer. The picks come in different weights, 12, 14, 20 and 22 ounces, with handles of wood or all metal, and with two different heads, one with a pick-like end, and the other a chisel edge. Both hammer heads are useful and a wooden *versus* an all metal handle are matters of personal preference. A wooden handle will absorb more shock from hammering on rock than will a metal one. The 22 ounce hammer is more useful for an adult — the lighter one more useful for youngsters.

Some collectors use other hammers, especially those known as engineers', stake, tile setters' and brick layer hammers. Next in importance is the sledge hammer. For most moderately heavy work a small sledge with a head weighing 5 to 8 pounds will do nicely. For the bigger jobs, a 10 to 12 pound sledge hammer is

Figs. 5-1, 5-2. The assortment of important tools and equipment at left should always be taken on field trips, for lack of one of them might mean failure to get a specimen. Perhaps the most-neglected aspect of collecting is in aids for wrapping: at right, the recommended heavy cartons, egg boxes, newspapers, paper and burlap bags, tissue and knapsack, are made ready for packing in the car.

used, preferably one with a hammer end and opposed chisel-edge head. Such a hammer can be used as a wedge on occasion by being struck with the smaller sledge. The heavy hammers are used for breaking up large boulders or splitting open veins or seams in walls or loose chunks of rock.

The other two large tools, the broad-faced shovel and the ditch-digging pick are used for digging purposes, such as digging into a dump or opening up the ground at a likely looking spot. Their use is dictated by the locality and prevailing conditions.

The large chisel-like tool, called a moil, and the carpenter's bar are very important. They will do the job of prying apart rocks along fractures, splitting off large sections of pocket walls and a variety of other jobs.

A good assortment of cold chisels is essential. They are used to split smaller rocks and work within pockets, vugs and veins. Carrying three or four chisels varying from $\frac{1}{2}$" to 1" in blade size is recommended. A pair of jaw-like nippers or pliers is used for nipping off sharp edges.

The cold chisels and similar edged tools should be kept sharp and it is a good idea to carry an extra handle for the wooden-handled tools.

Surprisingly enough, small garden hand cultivators or similar hand tools may prove very useful on a small dump, if the material is finely broken, or for turning up crystals in loose dirt areas where they are eroded out of rock matrix.

Another tool that may prove useful is a "pocket scraper," which is simply a short length of steel with a flattened right-angle end about 1" in length. This is used to hook loose crystals or groups located in pockets that cannot be reached by other means. Pockets in hard rock may have loose crystals in them that defy the collector unless such a tool is available.

An inexpensive 10 to 14 power (magnification) hand lens is a valuable asset to have tucked in your pocket for field work. It enables you to examine specimens under magnification and helps identify many materials. Hand lenses get lost easily in the field so attach it to a belt loop by a length of strong cord or leather thong. Most experienced mineral collectors are never without their hand lenses, whether collecting in the field or visiting someone else's collection!

Another small pocket aid having a dual use is a compass. Its main use is as an orientation and direction finder, and if long hikes over unknown, forested or hilly country are undertaken, it should be used. Its secondary purpose is to test for magnetism. Suspected minerals may be held near the compass and if they cause needle deflection, it indicates some degree of magnetism.

No two crystal localities will provide the same collecting conditions. An experienced collector may work for hours to gather a few fine specimens. Skill in specimen removal is the product of patence and experience, and this can only be obtained in the field.

As previously mentioned, crystals are usually found lining cracks in the rock, in pockets or vugs. Collecting becomes a matter of removing the crystals themselves, or sections of the pocket or vein walls with the crystals attached to some portion of the matrix. Experienced Swiss collectors, who often found spacious vugs in the Alps lined with large quartz crystals, used a technique called "packing" to protect the crystals during removal. Removing single crystals or groups often caused specimens to fall against one another within the vugs resulting in fracturing and feathering of the crystals. Consequently, vugs would be packed with sand, soil, or forest duff, forming a cushion for the specimens as they were split from the sides of the caivty.

In the case of the Swiss vugs, proper tools for collecting are absolutely essential, long moils and small sledge hammers being the most important.

Packing Specimens in the Field: Coupled with the importance of proper tooling is the factor of proper packing or packaging of specimens after they have been removed. For the most part, it is impractical and time consuming to stop for detailed inspection, cleaning and preparation of specimens as they are collected in the field. However, they must be properly wrapped or protected until they can be transported home for inspection and preparation for the collection.

Many fine specimens have been collected and then ruined by careless transportation down a mountainside to the car. Specimens or crystals should always be wrapped or similarly protected. Wrappng specimens in newspaper and then packing them in a knapsack, gunnysack, basket or box is recommended. The little extra time it takes to wrap specimens in the field is well worth the effort. Fig. 5-2 shows some simple aids to packing and wrapping specimens in the field. Here are a few tips regarding preparations for collecting and wrapping which have proved effective.

1. Rolled newspapers prepared in half-sheets.
2. Rolled, used tissue paper.
3. Package of used small paper bags.
4. Cleansing tissue.
5. Egg cartons.
6. Knapsack.
7. Extra burlap (gunny) sacks.

With the exception of the cleansing tissue and the knapsack, these materials cost nothing. The entire kit of materials illustrated weighs only two pounds, so it adds little to the few tools you may carry to a site. Small tools, such as the cold chisels, will pack easily in the knapsack pockets.

A properly adjusted knapsack is by far the best means of carrying a load of specmens a long distance back to the car. An extra gunny sack or two should always be carried just in case you make a major find.

Newspaper is the most satisfactory wrapping medium for most specimens. Rolled newspapers are most easily carried. In preparing for collecting, lay several thicknesses of newspaper out on the floor completely unfolded to their largest dimension. Cut or tear the paper along the main fold and restack. This gives you single half-sheets, the most usable size for wrapping. Roll up the sheets and put an elastic around them. When in the field simply unroll the sheets, place a rock on the pile, and wrap. This prevents clumsy unfolding of quartered newspapers and minimizes the chance of their blowing away from you during the process.

Learn to wrap specimens by folding the newspaper as many times as possible over that side of the specmen containing the crystals as it is wrapped. This simply puts more protective cushion around the crystal area.

If the specimens collected are very fragile or delicate, additional steps should be taken to assure their safe transportation. An initial wrapping in soft tissue paper or bathroom cleansing tissue is advisable. Tissue paper that comes in suitboxes and gift parcels, flattened and rolled up, makes an inexpensive source of such packing.

For very delicate and small specimens, an egg carton is a handy item. Wrapping the specimen in tissue and placing it in one of the box's partitions provides good protection.

Another wrapping medium is the paper bag, especially the small ones almost every household throws out. Saving these during the months when you cannot collect makes wrapping very easy. A stack of several dozen flattened small paper bags, tied together with string or strong rubber band, takes up little room in a knapsack and along with the roll of newspaper and tissue, involves very little weight.

A few small, sturdy cardboard cartons filled with cotton are good for those extra-special and very delicate specimens you may acquire on a field trip. Four or five such boxes kept in the car will do much to facilitate their safe arrival.

FIELD NOTES OR LABELS: It is a good idea to get in the habit of making detailed field notes or labels for specimens as they are collected. An inexpensive pocket notebook or small pad of paper will suffice. Some collectors write field labels for each specimen and enclose it with the item as it is wrapped in the field. This is not always necessary, and making a series of notations in a pocket notebook regarding specimens as they are found is usually sufficient.

As a collection grows and you take more frequent field trips, it is almost impossible to remember all the important information regarding the various localities visited. Enclosing a field label with each specimen or group of specimens from the same locality refreshes your memory. A field label or note should contain the name of the specimen (if you are able to identify it), the exact and detailed locality (county, nearest town, name of mine, quarry, etc.), date collected, and any related remarks applicable to observations of the locality in general or the specimen itself. Also the names and addresses of any person responsible for your collecting on a site should be included. A brief letter of thanks upon your return home does much to pave the way for some other collector, or for a return visit on your part.

CLEANING SPECIMENS: Frequent washing of collection specimens should be avoided, but there are times when washing will greatly improve a specimen's appearance. Apparently worthless specimens may surprise you after they are soaked, washed and cleaned. Specimens found in cracks, pockets, or loose on the ground may be mud-caked, dirty, or stained. Specimens obtained from old collections may need washing to remove grime and dust. In many cases washing in lukewarm soap and water using a toothbrush, paintbrush, or scrub brush, can reveal hidden beauty.

Specimens should not be scrubbed with scouring powder unless the specimen is harder than the powder. Scrubbing a group of selenites with a scouring agent could ruin them, since selenite is very soft. The same powder could be employed with safety on a quartz crystal. *Never* wash specimens in very cold or very hot water unless you experiment with an unwanted sample first. Crystals suddenly plunged into cold or hot water may shatter or fracture.

Be sure the specimen is *not* soluble in water. If halite or hanksite is washed, the collector may find himself with nothing but a bucket of saline solution!

Certain stains on mineral specimens may be removed by chemical action. It is best to consult some other collector, museum curator, or professional mineral dealer regarding the possible correct method to use for the particular mineral or combination of minerals involved. A saturated solution of warm oxalic acid, or weak hydrochloric will remove many stains, as will the new "safety acids" such as SR-1. However, the use of too concentrated acid may etch crystal faces or even run an acid-soluble mineral.

Some specimens may call for the dissolving away of one mineral in order to expose another. An example is the benitoite, neptunite, and joaguinite crystals embedded in massive natrolite from the famed San Benito County, California, locality. Specimens are immersed in acid which dissolves the surrounding natrolite and exposes the attractive blue, black, and brown crystals.

TRIMMING OR DRESSING: There are two reasons for trimming or dressng specimens. One is to reduce a specimen to a particular collection size and the other is to eliminate extraneous and unattractive portions that detract from a specimen. Skillful trimming can greatly improve many specimens.

Trimming is tricky! Even the experts break specimens. It is best to experiment on specimens of poor quality first. Different types of rock will break in a variety of ways. Mineral dealers and college geology departments have heavy duty rock-trimmers, a screw press-like device with two vertically opposed large chisel blades. The private collector will find a chisel-edged trimming hammer, nippers, pliers, small cold chisels and dentist tools valuable for trimming and dressing specimens.

Trimmng can be done with the specimen held in the hand, placed on the ground, or on a block of soft wood. If specimens are hand held, be sure to cushion the specimen in a heavy glove or several thicknesses of cloth. Not only does this afford better resiliency for trimming, but it also protects your hand from sharp slivers of rock. Eye glasses or protective eye shield should be worn during trimming and dressing.

Occasionally a fine group of crystals will surmount a disproportionate and unattractive rock matrix. If the unwanted section cannot be split or trimmed off by other methods, it may be sawed off on a diamond saw. It is important that the lubricant used in the saw does not stain the crystals. Most modern lubricants are easily cleaned from specimens by luke warm water and soap.

Dentist tools can be particularly valuable in dressing specimens. They are used for the removal of caked mud or other foreign matter situated deep between crystals, or for removing crusts or coatings from some crystal faces. Care must be taken not to scratch a specimen. In some cases a collector might spend hours dressing a particular specimen. This would be an infrequent occurence, but some excellent specimens have been prepared only by the painstaking use of such small tools.

DUPLICATE SPECIMENS: Good duplicate specimens make fine trading items. It is almost impossible for a collector to visit all the good collecting localities in the United States, and even then no assurance could be given of having good luck at all of them. Trading duplicate specimens with other collectors is one of the best ways to broaden and improve a collection. Many collectors have really established their collections by trading quality "local" material in all parts of the globe with other collectors, dealers and museums.

SPECIMEN STABILITY AND PROTECTION: A popular misconception about crystal specimens deals with their relative stability. Crystals are often considered in the same general classificaton as rocks, with the implication that crystals possess the same durable characteristics associated with rocks. This is not necessarily true.

One of the most overlooked aspects of mineral collecting is the care and preservation of the specimens once they are obtained. Lack of information in this regard has resulted in ruined specimens. The great majority of crystal specimens may be considered *stable*. In other words, they resist alteration and change and with proper care may be handled and examined with ease. A number of crystallized minerals appear delicate and therefore receive careful handling based purely on physical appearance. But this subject is much deeper than appears on the surface. There are both common and rare minerals which show or undergo various physical and chemical changes once they are removed from their environment and located in a collection. Such specimens are referred to as being *unstable*.

From the standpoint of geological time, all mineral specimens (and rocks) must be considered as having a life period measured in hundreds, thousands, or millions of years. Eventually everything alters or changes; nothing in our physical world is completely static or permanent. All collectors find specimens that are in various stages of alteration, and change can take place within a period of a few hours, months or years.

A positive evidence of change is apparent in the distinctive odor surrounding most mineral collections. Although slight and not offensive in the least, the odor of a collection is technically a combination of gases being released by certain specimens as they react to air and moisture.

What can affect a specimen? Depending upon the mineral in question, and sometimes on its particular locality, different conditions can affect its relative stability in varying degrees. Slow changes can be produced merely by air circulating around the specimen, which produces an environment of oxidation. The finest piece of steel pipe will crumble into dust or rust away within a few years if left out-of-doors exposed to only two factors — air and moisture. The same process, at a much slower rate, may operate on specimens. Examples of minerals that oxidize or hydrate and can noticeably change in a collection are pyrite, native copper, realgar, hanksite and marcasite. Pyrite and copper can assume a dull lusterless tarnish, realgar turns to an orange-yellow powder, hanksite gains a rough white crust, and some marcasite can disintegrate to a black powder. Such changes are commonplace and are the product of various chemical elements in the mineral reacting to, or combining with, air. They are also affected by temperature or condition of the air, whether it is warm, cool, moist or dry. Some specimens actually dry out, and fine opal specimens (an amorphous form of quartz) have been known to shatter in cabinet drawers for this reason. Such minerals are referred to as *hydrous,* which means that water is one of the specimen's basic components.

A few rare minerals must be kept under refrigeration, otherwise they alter almost immediately. The mineral mirabilite from Searles Lake, California, is a good example. Under normal room temperature conditions, this specimen actually dissolves in its own water of crystallization. Other minerals that exist as liquids at ordinary temperatures are water and mercury.

Another type of alteration has been mentioned in connection with physical properties in another chapter — the phenomenon of photosensitivity in which light affects a specimen. Examples are proustite and vivianite, both of which may alter radically if excessively exposed to light. In a few cases, usually confined to specific locality peculiarities, minerals have been known to simply fade in color as time passes. Certain brown and pink topaz, calcite and turquoise may fade.

In order to protect certain specimens a knowledge of the fundamentals of chemistry is helpful. Nothing can be done to prevent fading, except in the case of known photosensitive specimens which can be kept in boxes away from light. Steps can be taken to protect against other types of alteration, and most advanced collectors and mineral dealers are glad to furnish practical and specific information regarding certain minerals. Clean, bright specimens of pyrite and native copper can be somewhat protected by a thin transparent coating of plastic spray. Plastic spray or liquid is better than lacquer, shellac or varnish, which can markedly discolor a specimen. The same applies for various salts and borates. The principle involved is that of excluding air from actual contact with the specimen. Some collectors and museum curators use dehydrating agents, such as silica-gel, placed in small quantities in display cabinets. These chemicals absorb and stabilize air moisture and minimize alteration.

Chapter 6   Specimens Size and Collection Types

A common misconception is that mineral collections are in effect "all the same." This is quite inaccurate for, although the majority of mineral collections are of a general or comprehensive nature, they can show astonishing variety, difference in type, emphasis, specialization and originality.

Collectors may choose to specialize in one particular phase of mineral collecting just as other hobbyists, such as philatelists, may specialize in collecting British colonial stamps, or the numismatist in old coins of Greece and Italy. The reason is simply interest, a phase which strikes the fancy of the collector, or perhaps the realization that specialization may make for a more meaningful and valuable collection.

There are several variables involved in typing mineral collections but they are essentially based on a combination of two considerations: (1) the size of the specimens collected, and (2) the major emphasis or specialization, if any, of the collection. These are of additional importance if a collector wishes to exhibit his materials in competition at local, state or national shows and conventions.

The following discussion of specimen and collection types covers those most often found among both amateur and advanced mineral collectors. Despite the variety mentioned, you may be sure that someone may have a particular interest in collecting a different size, with a specialization emphasis on some other aspect of mineralogy than those mentioned. These are, however, more-or-less standardized sizes and types accepted by the various mineralogical societies, and many are the product of experience and convenience as the hobby of collecting crystallized minerals has grown through the years in the United States and abroad.

THE IMPORTANCE OF SPECIMEN SIZE: Although it may seem relatively unimportant at the outset, most collectors soon find practical advantages in focusing their collecting within the bounds of an approximate size limitation. There are several sound reasons for this. The most important is the over-all professional appearance and neat format of a collection in which the specimens selected show an approximate size limitation. It is certainly more attractive than a hodgepodge of specimens varying from microscopic crystals of joaquinite to a sixty-five pound quartz crystal.

Another key factor in size consideration is that of the price of mineral specimens. Most collectors, at one time or another, purchase specimens. It is simply impossible to find all the specimens you would like to have and you cannot always trade for what you want. Therefore, cost is a very realistic factor.

Small specimens naturally cost less than large ones, and when it comes to purchasing rare and exceptionally fine specimens the average pocketbook might be able to afford a 1"x1" or a 2"x3" whereas a large cabinet size specimen might be prohibitive for most collectors.

Another point that is important, and often grossly misunderstood by beginning mineral collectors, is that you do not have to have a large specimen, or large crystals, to have a good quality specimen! Actually, the finest examples of crystallization are found in the micromount and smaller-sized specimens, since the most perfect natural crystals are usually small.

The selection of a mineral size to collect or emphasize is purely a personal matter. Most beginning collectors start with a variety of sizes, since this is what they find in the field, and then gradually swing around to a favorite size. It should also be stated than many collectors collect specimens in a number of sizes simultaneously.

The size of collection specimens ranges from those that have to be studied with a microscope to those of huge size suitable only for display in a museum. The most

Figs. 6-1, 6-2. Specimens at left are micromounts in typical housings, from the large Benjamin Chromy Collection. At right, "thumbnails" (1"x1"), including azurite and cerussite from Arizona at upper left, galena from Kansas at upper right, marcasite from Rumania at lower left and endlichite from Mexico, lower right. Note comparison of sizes with the white one-inch cubes placed in the photos' centers.

well-established and recognized size categories are as follows, beginning with the smallest.

MICROMOUNTS: During the last few years this has been the most rapidly growing newer aspect of the crystallized mineral hobby. Micromounts are tiny specimens, recognized officially by most mineralogical societies as single crystals or groups, with or without matrix, of any size that can be mounted within a standard "micromount box" not larger than $\frac{7}{8}$"x$\frac{7}{8}$". Such boxes, or individual specimen housings, are of cardboard, black or clear plastic and are handled by most mineral dealers or scientific supply firms. See Fig. 6-1. Micromounts are also often displayed on small cubes or bases of styrofoam within a divided box.

In order to observe micromounts, as the name implies, they must be viewed with the aid of some form of magnification. There are some advantages to micromounts. First, they can be quite spectacular in terms of crystal quality since very small microscopic crystals tend to be the most perfectly formed. Since the specimens are small, a good many hundred specimens can be housed in a small area, a factor which is important to a hobbyist who may have limited specimen display or storage space. Another important point which is common to all small specimens is that of initial cost, if they are to be purchased. Such tiny specimens cost little compared with larger pieces. Also, micromount size may allow the collector to afford rare crystallized specimens. Often a mere chip or fragment from a large crystal group constitutes a fine micromount.

Boxes containing micromounts are usually stored in groups in larger boxes, or in small drawer cabinets such as those housing dental tools or spool cabinets.

"THUMBNAILS" OR 1"x1"s: As the classification indicates, these are small specimens about 1"x1" or slightly smaller. This is a popular Dana Collection size. The advantages of such a size are shared with micromounts. In other words, good quality in small specimens, ease of storage and display, availability of many varieties at comparatively low cost if purchase is contemplated, are important. In addition, no microscope is necessarily needed to observe the mineral properties, and a well selected collection of 1"x1"s can be quite attractive. This size is illustrated in Fig. 6-2.

Figs. 6-3, 6-4. Miniatures at left (specimens up to size 1½"x1½"x2") are wulfenite from Arizona at upper left, smithsonite-coated cerussite crystal from S.W. Africa at upper right, vanadinite from Arizona at lower left, and calcite and cuprian austinite from Mexico. In photo at right are cabinet specimens (from 3"x4" to 4"x5"), showing Arizona wulfenite at upper left, Russian topaz at upper right, Utah pyrite at lower left, and Kansas galena and sphalerite. One-inch cubes in photo centers are for specimen-size comparison.

Fig. 6-5. Large cabinet specimens seen with one-inch cube: Uruguay amethyst at upper left, Missouri calcite at upper right, Illinois fluorite and quartz at lower left, and Kansas galena and sphalerite.

Fig. 6-6. Classic "museum size" specimen of cassiterite from Araca, Bolivia, measures about 20" in length. (Minerals Unlimited photo)

MINIATURES: This refers to small specimens up to, but not larger than, 1½"x1½"x2". This is a very popular size embodying all of the advantages noted for micromounts and 1"x1"s, plus the fact that a fine miniature of this size can be very showy and attractive. See Fig. 6-3. Miniatures are usually displayed in small glass cases or housed in individual box trays in drawer cabinets. From the standpoint of ease of storage with good display potential for most home collections, the miniature is ideal.

CABINET SPECIMENS: This term has several implications, but in general refers to specimens of sufficient size to be housed and displayed to good advantage in a glass case or cabinet. This means specimens up to handsize, or approximately 4"x5", as appear in Fig. 6-4. For the average collector interested in display, this is the best size, but such specimens may be relatively expensive. Also, large specimens in excellent condition and of high quality are harder to find than the smaller sizes. The term "cabinet specimen" is certainly not specifically limited to the mentioned 4"x5", and many collectors specialize in smaller-size cabinet specimens which also display well, particularly a size approximately 2"x3". For standard study and reference collections, most American museums, colleges and university geology departments use a 3"x4" specimen size.

LARGE CABINET SPECIMENS: As the classification indicates, these are larger specimens suitable for display in cabinets either as the main collection-size unit or as specimens forming the back rows on shelves of smaller cabinet-size specimens. Specimens in this size are usually about 5"x10" or larger, as illustrated in Fig. 6-5. Many museums emphasize this size.

MUSEUM SPECIMENS: Although museums display a variety of sizes according to available space and the particular type of display involved, the term "museum specimen" refers to very large specimens weighing perhaps several hundreds of pounds. Such huge specimens seldom display well in a private collection unless ample room is provided for proper viewing. A museum-size specimen of cassiterite is illustrated in Fig. 6-6. When huge specimens are contained in a private home collection along with smaller units, the large specimens sometimes overshadow and dominate the collection and actually detract from the collection as a whole.

THE IMPORTANCE OF LOCALITIES: A factor which is involved in any type of mineral collection, irrespective of the size specimen involved or the particular specialization, if any, is that of the mineral specimen's source — its locality or where it was found.

To the mineral collector, the locality of a specimen can be almost as important as the specimen itself. As mineral specimens differ one from the other, so do different locality characteristics for the same mineral.

There are dozens of notable mineral specimen localities wherein the particular form of crystallization is highly distinctive. To an experienced collector, localities are as easily recognizable as are the specimens themselves. Such ability to recognize localities is the product of experience.

COLLECTION TYPES AND SPECIALIZATIONS: The following types or emphasis classifications are maintained in a variety of sizes according to the particular desires of the individual collector. There is ample room for specialization within the mineral collecting hobby, and most of the major mineralogical conventions and shows in the United States provide specific competitive exhibits within the following general classification framework. Sometimes specific size limitations are enforced when collections are exhibited in competition.

THE GENERAL COLLECTION: This is by far the most common type of collection, especially with beginning collectors. The beginner will also learn more about minerals and crystals from such a collection. As the classification indicates, this consists of a wide variety of mineral materials and often includes related materials such as rocks, agates and fossils.

General collections may or may not be keyed to particular size, but any size or combination of sizes is applicable. Many experienced collectors may maintain a general collection but have a certain division or divisions within it devoted to particular specializations.

THE DANA COLLECTION: A Dana collection involves the attempt on the part of a collector to obtain as many as possible of the over 4,000 different mineral species listed in Dana's *System of Mineralogy*. Those who work on a Dana collection usually confine collecting to smaller specimens, 1"x1" or 2"x2", since many items will have to be purchased and many Dana minerals occur only in very small specimens, powdery forms, or minute traces in matrix. A complete Dana collection is almost an impossiblity, since many minerals are limited to a mere handful of specimens from one locality. However, it is possible to obtain several hundred different mineral species and the Dana collection is unquestionably the most scientific of all collection types. A Dana collection also implies an emphasis on rare minerals.

Another aspect of a Dana collection is the obtaining of as many as possible of the listed and famous localities for mineral species appearing in the Dana texts. Localities can be just as important as the mineral itself, and the phrase "an old Dana locality" has added meaning to the Dana or other advanced collector.

SPECIALIZED COLLECTIONS: Beyond the General and Dana collections there are several well established specialized collection types. Following are a few typical examples of mineral collection emphasis and specalization. Such concentrations enable the collector to build a collection which is both unique and distinctive and which provides an opportunity to obtain better coverage of a smaller field.

ONE MINERAL: Many collectors specialize in one mineral only from a number of localities, such as quartz, calcite, barite, wulfenite, garnet, tourmaline, fluorite or sulfur. Collections of this type can be extremely interesting, especially if one of the more common minerals is selected, such as quartz or calcite. Despite their abundance, these minerals occur in a wide variety of forms with hundreds of unique and highly distinctive localities involved. A fine collection of fluorite, barite or calcite can be breathtakingly beautiful.

ONE MINERAL FAMILY: In this case the specialization is limited to specimens within a single mineralogical family, such as lead minerals, iron minerals, or copper minerals. Again, such collections can be very attractive.

ONE GENERAL CHEMICAL CLASSIFICATION OR ELEMENT GROUP: This is a slightly different approach to the mineral-family type of collection, with the exception that the emphasis is placed upon chemical classification; i.e., a collection of oxides, sulfates, silicates, carbonates, phosphates, etc. This is also modified to emphasize elements as in an element-group type of collection.

ONE COUNTRY, STATE OR COUNTY: This is a broad regional classification and examples might be collections of minerals of Switzerland, South Africa, New Jersey, Madagascar, Brazil or Montana. A narrower focus might be upon the minerals of Calaveras County, California, or Hardin County, Illinois.

ONE MINING DISTRICT, MINE OR LOCALITY: Often called a "suite" collection, this is a popular type in Europe, and many old European collections placed great emphasis upon an individual locality. Such collections might contain dozens of specimens of the same mineral from one special locality, such as smoky quartz from Fellital, Canton Uri, Switzerland.

Collections focused on a specific locality or mining district might be as follows: minerals of the Castle Dome Mining District, Yuma County, Arizona; the Tsumeb District, South West Africa; Tri-State District, U.S.A.; Franklin Furnace District, New Jersey; or the Magdalena Mining District, Soccoro County, New Mexico.

A particular suite of specimens from an individual mine could be: minerals from the Mammoth St. Anthony Mine, Tiger, Pinal County, Arizona; the Crestmore Quarry, Riverside County, California; or from the Bigrigg Mine, Cumberland, England.

SINGLE CRYSTALS: A very popular specialization is that of the single crystal. Collecting is usually confined to smaller specimens although a few hobbyists have specialized in large single crystals. Such a collection is of high value and interest to the person seriously interested in crystallography, since single mineral crystals are the easiest to study and measure.

GEM STONE MINERALS: This classification cuts across many others, but some collectors have specialized in gem stone crystals and gem stone minerals. Such a collection can be very handsome but costly to maintain and enlarge when in the advanced stage of development. There are a number of magnificant privately owned gem stone mineral collections.

PSEUDOMORPHS AND UNUSUAL MINERAL FORMS: The phenomenon of pseudomorphic mineral specimens is covered in another section of this book. Pseudomorphs make an interesting, if somewhat limited, specialization. Such a collection is often coupled with other odd and unusual forms such as twins, distortions and unique combinations with the emphasis on a mineral's physical structure.

SHOWY OR ATTRACTIVE MINERALS: A number of collectors specialize in collecting and displaying only mineral specimens which are well crystallized, attractive and showy. Of the hundreds of known minerals, well over half of them, even if found in well crystallized forms, are dull and comparatively unattractive.

ORES AND RELATED MINERALS: Although not particularly within the crystallized mineral field, collectors may become interested in various ores, an aspect of petrography, and their associated minerals. A collection may consist of ore samples from mines coupled with crystallized minerals occurring within various ore zones.

MINERALS PERSONALLY COLLECTED: Some collectors place major emphasis upon displaying only specimens they have personally collected. This is, of course, a type of collection that is very rewarding from a personal standpoint, and a type which commands respect if of fine quality.

Chapter 7   Housing, Cataloging and Labeling

Just as important as actually collecting specimens are the factors of cataloging and labeling them accurately, and housing or displaying them in an attractive and practical manner.

HOUSING THE COLLECTION: It is usually a problem for the beginning private collector to decide upon the most satisfactory method of display that is attractive, within home space limitations, of moderate cost, giving some protection against dust, and allows him to organize his collection in whatever way he chooses.

Before considering any type of display or housing, keep in mind that it is quite possible for a mineral collection literally to run a family out of house and home. A good method of display, which provides adequate expansion space and is limited to well chosen specimens, will do much to prevent such a condition. A neat, well displayed collection adds immeasurably to organization and enjoyment in viewing. Quantity is no substitute for quality.

Dust is the constant enemy of museums and collections of all types. Nothing is really dust-proof unless hermetically sealed; but there are display methods which reduce or minimize this problem. These, essentially, are the four methods of housing or display.

1. Specimen boxes
2. Open shelving
3. Glass cases
4. Drawer cabinets

SPECIMEN OR COLLECTION BOXES: These are individual flat boxes with hinged or removable lids obtainable in a variety of sizes and depths with or without dividing partitions. See Fig. 7-1. They are made of wood or heavy pebble-finished cardboard and may be purchased from a number of mineral dealers or scientific supply houses. They may be made at home but it requires fairly exacting carpentry. Specimen boxes are lightweight and portable.

This method of housing is effective and particularly well adapted to the smaller size specimens such as the micromount up to 2"x2". A great many specimens may be stored in a small area (a shelf or bookcase) and are well protected from dust.

Pre-fabricated specimen boxes are available through dealers with 100, 75, 50, 24, 18 and 10 individual compartments for specimens. Most collectors and mineral societies have more-or-less adopted as a standard box for displaying "thumbnails" a rectangular box containing 50 one-inch compartments or divisions. Many collectors use specimen boxes containing no built-in partitions, but fill the box with specimen trays of the appropriate size accommodating their specimens.

Sometimes specimen boxes are used with unusual methods of mounting in which wire expansion holders secure the specimens above a styrofoam base. Specimens have been mounted atop clear plastic golf tees and on home-designed plastic bases of many types.

Some collectors use a special glass-topped box called a Riker Mount for holding small crystals. The specimens are matted against a white cotton background. Riker Mounts come in a variety of sizes. Similar to these mounts in purpose are button frames, a standardized method of displaying buttons for those who follow that hobby. Button frames are glass-faced frames or boxes, and specimens are affixed to a sheet of heavy pebble board at the back of the frame. These have a depth of one to two inches. Both button frames and Riker Mounts have a variety of

Fig. 7-1. Specimen boxes are useful for housing small specimens. They are available in various sizes with or without partitions, and they may be fitted with individual trays to suit collector's need.

uses other than for minerals, and will accept butterflies, arrowheads, small fossils, leaves or other small objects.

OPEN SHELVING: This is the most inexpensive method of display in terms of the number of square feet of exhibit space available at low cost. However, open shelving is subjected to the maxmum accumulation of dust. If open shelving is used it is advantageous to mount the shelves so they can be adjusted. Track for such mounting is relatively inexpensive and allows you to adjust for shelves holding large or small specimens. Also, if shelving is painted, it should be stained or painted with a non-glossy or satin finish paint. Shelving having a glossy painted surface may detract from specimens.

Specimens are usually simply arranged in an attractive manner by placing them directly on the shelf surface. Another method, which is also applicable to glass cases, is to place the specimen on a "base" or stand of wood, styrofoam or plastic.

GLASS CASES: Glass cases, in a great variety of sizes, types, and shapes are commonly used by mineral collectors. Their advantage is essentially one of display since minerals show up well in a glass case, especially if lighted, and are well protected from dust.

The types of cases used by private collectors vary from custom built installations such as in Fig. 7-2 to more portable units as illustrated in the two cases in Fig. 7-3 housing cabinet specimens in the author's collection. New glass cases may, of course, be purchased from showcase companies, and some manufacturers produce very practical units which can be stacked one atop the other as the collection grows. Most collectors, however, obtain used glass cases to keep the cost low.

Fig. 7-2. Attractive, well-illuminated, custom-designed wall cases display the Ronald Sharps Collection in his California residence.

There are several sources for second-hand glass cases, depending on the type and size the collector wants. Showcase companies usually have an ample stock of used units, as do some store fixture and restaurant equipment firms. They may also be obtained at auctions of store fixtures, fire sales, and "going-out-of-business" sales. Used cases may be in need of refinishing or minor glass replacement, but in any event, the savings involved are well worth considering.

Glass cases may be illuminated by either incandescent bulbs or fluorescent tubes, however, the latter may disguise a mineral's true color. Showcase vendors and electrical contracting and appliance stores are sources for these items. Some

Fig. 7-3, 7-4. A portion of the author's collection is seen in glass cases, left, containing crystallized minerals of cabinet sizes; and at right, in specially-designed drawer-cabinets (used for miniatures) whose capacity is over 1,900 specimens. Note wide-swinging lamp.

collectors prefer a glass case that has a mirror backing, or even mirror shelves. Others find mirrored surfaces confusing and feel they may detract from specimens exhibited. It is simply a matter of personal preference. Generally, more collectors use clear glass cases with clear glass shelves.

For purposes of displaying specimens in competition or for participating in an exhibition, the various mneralogical societies or federations have generally adopted a particular exhibition case format and size limitation. Exhibits are usually confined to a glass-fronted shelf case 2x4 feet with maximum space of twelve square feet, or a display consisting of two of these cases side by side with a maximum of twenty-four square feet. Most societies have developed excellent plans for building such exhibition cases including lighting, at low cost. The exhibit chairman of a society will gladly furnish details to interested collectors.

DRAWER CABINETS: While the glass case is outstanding in terms of beautiful and impressive mineral display, drawer cabinets excel in providing a precise, neat organization of a collection within a series of drawers. Drawers are more dust proof than glass cases, they take up less room and can accommodate many hundreds of specimens particularly of the small sizes.

Because of the flexibility of drawers, a collector may organize his collection by whatever system appeals to him. For example, the basic divisions of the Dana System of Mineralogy may be followed in drawer sequence by the Dana collector. Perhaps specializations such as all copper minerals are placed in one drawer, calcite crystals in another, and so on.

Drawer cabinets can be made by a collector, made to order, or purchased commercially. Fig. 7-4 shows a pair of seventeen-drawer cabinets which were made to order and which house a portion of the author's collection of specimens up to the 2"x2" size. Other sources rest with the showcase, store fixture and restaurant supply firms where drawer cabinets may be sold for other purposes. Items such as map, spool, small tool, hosiery, or dentist tool drawer cabinets often lend themselves to mineral use.

Metal drawer cabinets of varying dimensions are also handled by some mineral and natural science supply houses. A relatively inexpensive and excellent drawer housing for small specimens may be found in some 35mm slide and filmstrip drawer files. These may be carried by audio visual education dealers who sell the units primarily to schools. If you purchase a drawer cabinet, be sure you either anticipate future growth of your collection at the time you purchase, or that the unit you buy can be matched or added to at a later date.

Drawer cabinets are almost always used in conjunction with some size "specimen tray" or styrofoam base in or on which specimens are placed, the collector's label underneath. This is standard procedure followed by hundreds of collectors. A swinging adjustable lamp should be located above the drawer area so that specimens may be viewed with ease. They are available in standard bulb or fluorescent fixtures.

SPECIMEN TRAYS: The term "specimen tray" refers to a simple tray-like pasteboard box, usually lined in either black or white glossy paper. They can be purchased commercially, or the collector may have access to some free source of small boxes which serve the purpose just as well. For example, the boxes housing slide covers used in microscope work in hospitals and universities make fine specimen trays for very small specimens. Collectors may have friends in a business where some small box is emptied and thrown out rather consistently, which may mean an inexpensive source of supply.

Specimen trays come in a variety of sizes (Fig. 7-5), including some with a shelf on one end for affixing a label. They are of greatest use within the drawer type cabinets. A tip regarding their use in a drawer is to cut strips of cardboard the width of the drawer, slightly lower than the tray height, and slip these in as runners between rows of trays. This enables you to slide them along easily as adjustments are made for changing specimen sequence or adding new specimens. A drawer of miniature specimens housed in specimen trays is shown in Fig. 7-6. Cardboard runners extend horizontally between each row of specimens.

GLASS OR PLASTIC VIALS OR BOXES: Often, collectors wish to protect some particularly fragile or unstable small specimen. Plastic boxes and vials are available for this purpose, the only disadvantage being that they scratch easily. For this reason, some collectors prefer glass vials. Whatever the collector's choice, it is a good idea to settle on one or two consistent methods of housing such items, making the overall appearance of the collection much better. A drawer filled with good specimens in a conglomeration of different pill bottles, plastic boxes, etc., is distracting.

Small one-inch plastic boxes have become very popular with micromount collectors, and such tiny specimens can be mounted attractively in these units.

SPECIMEN BASES: Specimen bases are preferred by some collectors in their own home display, and often appear in museum displays and collectors' exhibits at mineralogical shows and conferences. Specimen bases may add a professional touch to a collection. Specimens are placed atop bases made of wood, styrofoam or clear plastic. They are usually square or rectangular, of varying dimensions according to the size specimen displayed, and from $\frac{1}{2}$" to 1" thick. The front or leading edge of a base is usually bevelled and a label is placed in this position.

Bases are almost always made by collectors themselves, although they can be purchased. If made of wood they should be painted a flat black or off-white. Sometimes collectors may flock the bases with colored flocking.

Styrofoam has been used recently by many collectors and these bases, or the stock for them, are carried by some mineral dealers. Florists also may carry large blocks of this material. Specimens are pressed into styrofoam and oriented so their best sides face to the front. If a fragile specimen is involved, an outline of its base area is drawn on the material and the area pressed down by a pencil or knife until the proper seat has been formed for the specimen. Styrofoam mounting can be showy, inexpensive and adaptable to many display situations. The only danger might be in over-use in a display to a point where the base material overwhelms the specimens! This is a frequent display error made by collectors exhibiting in competition at mineralogical shows and conferences.

Clear plastic bases are almost always made by the collectors who use them, or are made to specifications. They can be very handsome and clever in square, rectangle, circle, half-circle and rhombohedral shapes. At best, commercially prepared clear plastic bases are expensive. Base materials should be used with discretion — a display should emphasize the minerals, not the bases.

RISERS: The effective display space on any shelf may be increased by the use of a simple 1:3 step riser. Risers may be made at home from plywood, boards, or even blocks of wood. Risers are usually painted, covered with cloth, or flocked, and they serve to elevate each row of specimens on a shelf above the one in front of it. The dimensions of a riser are determined by the general size of the case or shelf, the depth of shelves involved, and the size of specimens to be displayed. A $2\frac{1}{2}$" step is usually quite satisfactory.

DISPLAY TECHNIQUE: How you house or display your specimens is entirely a matter of personal preference, type and size specimen involved, home space for display, and budget limitations. Here are some pointers:

1. Keep your housing method within budget lmitations. Inexpensive open shelving will do nicely if the specimens are well arranged and kept dusted, and secondhand glass cases are easy to obtain.

2. Don't overcrowd shelves and cases. Putting too many specimens on a shelf or in a case is the greatest weakness of the hobbyist. Displaying a few choice items

Fig. 7-5. Here are some of the many different sizes and styles of specimen trays commonly in use for storage inside cabinet drawers.

Fig. 7-6. Drawer of miniature wulfenite specimens in author's collection. Each specimen in its tray has its label under it, and thin cardboard runners separate trays. Drawer is not filled too tightly.

Fig. 7-7. Well-organized and uncrowded calcium and barium specimens in Vonsen Collection. (California Academy of Sciences photo)

is better than displaying many poor ones. Also, overcrowding of a shelf defeats its purpose as an exhibit area.

3. Arrange specimens neatly, with attention to the interesting and artistic possibilities involved. Notice how other collectors display their specimens, especially in exhibit competition; as well as the techniques and format used by museums.

4. Pay attention to good balance in a display. Do not indiscriminately mix large, medium and small size specimens on shelves without attention to this factor. Place the larger specimens on the bottom, use them as the back row, or let them flank the collection by locating them on the ends of shelves.

5. Mineral specimens lend themselves to imaginative and artistic arrangement due to the variety of shapes and colors involved. Exhibits and displays built on the time-tested principal of symmetrical arrangement, in other words, specimens of the same size in the same positions on either side of a case, are fine; but don't feel bound to this tradition. A case or series of shelves can contain an attractive symmetrical arrangement, but also one that is more imaginative with emphasis on one side of the case, a staggered arrangement according to specimen size, or a host of other possibilities. A simple and neat arrangement improves any display. Although a museum exhibit, note the interesting and pleasing arrangement of the large specimens in Fig. 7-7.

6. Frequently, specimens may be difficult to stand up, or orient, in the position which shows them off to the best advantage. The aforementioned specimen bases

may help; however, many collectors do not care for them. In such instances a small wedge or scrap of styrofoam, wood, cork or wax may be positioned on the lower rear side of the specimen so that it sits correctly. Sometimes heavy wire, such as used in a coathanger, may be cut and bent into holders or easles to accommodate specimens. Plaster of paris may be used as a base or wedge, but this is somewhat time consuming and offers little advantage over more readily available materials.

7. When placed directly on a glass or wooden shelf, a specimen may move out of its arrangement position due to the vibrations caused by walking about in the room. Such specimens may be kept in place by putting them on a rubber band or small piece of inner tube. This gives enough non-skid surface to prevent "creeping."

8. Specimens should be kept dusted. A large soft bristle paintbrush from 3" to 4" wide is perhaps the handiest dusting tool for specimens.

CLASSIFYING THE COLLECTION: The meaning of classifying a collection should be clarified before discussing cataloging a collection. Unfortunately these two terms, classifying and cataloging, are easily misinterpreted. They do not mean the same thing.

Classifying is the scientific designation of where a given mineral belongs in relation to other minerals, in terms of its chemical composition. Cataloging is your own recording system for organizing and keeping track of your collection. There are not many ways of classifying a mineral collection, and the revised Dana system of classification is perhaps the best. When a mineral has been located in Dana, and its Dana number logged in the collector's catalog, it is automatically classified from the mineralogical standpoint, and this information has been obtained from the foremost source book of its kind.

The revised classification system of Dana is based on crystal chemistry, a precise study based on morphology, the examination of external crystallographic appearance, and on the internal molecular structure of minerals as determined by X-ray and chemical analysis. Crystallochemical classification is concerned with the correlation between these factors. Briefly, the revised Dana System classifies minerals first by *classes,* then by *types, groups* within a type, and finally by the *species* in a group or *series.* Each class, type, group, species or series, has a number. For example, galena carries the Dana number of 2611; this is in Class 2 (the Sulfides), it is in type 6 (AX type compounds), in the first group (the Galena Group) within this type, and the first species, or member of a series listed; therefore the number 2611.

Mastery of the Dana System requires considerable study and experience; however, this system is recommended because it is flexible, easy to follow, and allows new minerals to be added to the classification system without disrupting the organization. Serious mineral and crystal collectors usually note the Dana number in their catalog reference to the specimen in question, and some collectors place the Dana number, in addition to the personal collection number, on the specimen itself.

CATALOGING: Cataloging is perhaps the most important aspect of organizing and maintaining a mineral collection. All mineral collections should be cataloged in some way, and whatever way you choose, follow it *consistently.*

The scientific value of a collection is greatly enhanced by adequate cataloging and labeling — they go hand in hand. Remember, the locality is just as important as the specimen itself. Much can be learned about the geology or history of an old mining district from its specimens if the detailed locality data accompanies them. In addition, proper organization and identification will increase the monetary value of the collection. There have been many instances where old collections of fine specimens have been obtained by a dealer, private collector, or left to some college or university, and found almost useless through lack of name and locality data. Experienced collectors and mineralogists can identify the minerals, and can even recognize many of the specimen localities through familiarity with locality characteristics, but there is always a chance for error in this.

Cataloging adds a professional touch to a collection, a certain desirable scientific respectability. It is unwise for a collector to rely completely on memory for specimen data as his collection grows through the years. As previously mentioned, much of the information regarding the great variety of known minerals has been made available through the studious efforts of many serious amateur and professional collectors.

The *ledger* and *card catalog*, Fig. 7-8, are the most frequently used recording methods for cataloging. In either, entries are recorded and added in sequence, with each new specimen receiving a personal collection or accession number as it is added. The card file has some marked advantages over the ledger. If a specimen is eliminated, or replaced by a better one, a new card is inserted in place of the old one, instead of crossing out the original entry in the ledger. The card catalog is thus quite flexible.

By using a card catalog it is easy to establish a cross file, a method which can make the data on a large collection much more accessible. Although there are different types of cross-filing techniques, the most common for mineral collections is by personal collection number, as well as by specimen name. Two differently organized card files may be maintained containing identical or similar cards, one filed by numerical sequence, the other by mineral names in alphabetical order. In the first, specimens numbered 506, 507, 508 and 509 follow one another, but may represent pyrite, jamesonite, adamite and vanadinite. In the second, these minerals would be filed in their respective alphabetical order under p, j, a and v. Under "d" might be found datolite, diaspore, dioptase, dolomite and dufrenite. Several specimens of dolomite might be in the collection and their cards would be grouped together. Cross-filing, if desired, may be tailored to fit any type of collection and the special interest of the collector. However, the straight numerical sequence file is by far the most common.

There are specific items of data which should be recorded whether a card catalog or ledger is used. It is wise to make a habit of recording detailed data on the specimen as soon as possible after finding or purchasing it. Accuracy and completeness of data are important. The following items are usually found on professional museum entries and also in the best private mineral collections. (It is not always possible to obtain all the data sufficient to complete the entries, and some collectors may prefer to delete various items.)

1. The collection number.
2. The Dana number (optional among collectors).
3. The name of the mineral, or minerals constituting the specimen.
4. The chemical composition of the mineral (optional).
5. Detailed locality data insofar as it can be obtained.
6. Remarks, such as physical description of the specimen, date obtained, donor or collector, interesting history, or purchase price.
7. Date collected or acquired.

The Dana number and chemical composition are deleted by some collectors because they feel they have no need for this data; or because (if the chemical formula is to be studied) they will look it up in a mineralogy book anyway. However, a chemical description of a mineral may be included rather than the specific formula. For example the formula for the mineral wavellite is $Al_3(OH)_3(PO_4)_2 \cdot 5H_2O$ which may be stated as a "hydrated basic phosphate of aluminum." The Dana number is probably of more interest to the advanced and serious student of mineralogy. Some collectors in this category record the page number of Dana where the specimen

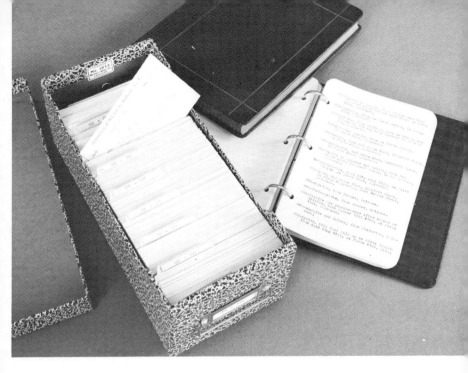

Fig. 7-8. The card catalog and ledger, either typed or handwritten, are most-used methods of keeping a collection's records. Card catalog is versatile, easily maintained. Many collectors use them both.

is described as well as the regular Dana number. For the mineral rutile such an entry would be *Dana No. 4511, Vol. 1, p. 554.*

Whenever possible, the catalog listing of a specimen should include as much data as possible in terms of a complete locality. For example, a catalog listing of "Wulfenite, Arizona" is obviously better than not knowing where the wulfenite came from at all. However, if possible, the specimen locality should be described in more detail, such as: Wulfenite, Collins Shaft, Mammoth St. Anthony Mine, Tiger, Pinal County, Arizona; or: Wulfenite, Old Glove Mine, near Amado, Santa Cruz County, Arizona. Whenever known, give the mine name, nearest town, county and state, or comparable data if from a foreign locality. Good detailed atlases of the world and the United States are useful for looking up localities. So is a collection of detailed road maps, especially those showing political subdivisions. You can do a lot of armchair traveling and learn quite a bit about geography by looking up mineral localities. Likewise, the volumes of Dana give considerable detail on notable localities for a particular mineral. Incidentally, the older volumes of Dana, (1877, 1898, 1922, 1932), contain many old localities not listed in later editions.

In describing specimens, collectors use the terms covered under the discussion of physical features. Learn how to describe minerals and use this skill in cataloging specimens. Following are some sample entries applicable to either a card catalog or ledger. They contain typical data on different specimens in the author's collection which specializes in crystallized minerals. Note that the approximate size of the specimen is always given (this can be done in either inches or centimeters) and sufficient descriptive data so the reader has a reasonably good idea of what the specimen looks like in terms of size, shape, color and quality. A six-inch clear plastic ruler is very handy for measuring specimens.

The degree of detail and format in these entries, which are on cards, is a matter of personal preference; some collectors would not include as much, others would include more.

Name: Aurichalcite   No. 579
Locality: Tintic Mining District, Juab County, Utah
Data: A carbonate-hydroxide of zinc and copper. Radiating velvety tufts of micro-acicular turquoise and silvery colored aurichalcite coating a pinkish rock matrix. Size: 4"x3"x2". Showy. Collected by A. L. Inglesby.

Name: Wavellite and Quartz   No. 1216
Locality: Llallagua, Bolivia, South America
Data: A hydrated basic phosphate of aluminum. Three spherical aggregates of pale lemon-yellow wavellite xls on slender transparent quartz xls. Micro acicular inclusions in quartz xls. Fine radial structure reverse side of group. Brilliant, undamaged, showy. Size: 1¾"x1¼"x¾". Purchased Oct. 1959: $5.00.

Name: Chalcopyrite, Calcite, and Dolomite on Quartz   No. 1046
Locality: Prinzenstein, St. Goar on the Rhine, Germany
Data: Small, very showy and brilliant typical xls of all minerals involved. Thimble-shaped specimen, size 1½"x1¼"x1". From Schneider Collection, University of Berlin. Trade, May 1957.

Name: Neptunite and benitoite   No. 1035
Locality: Headwaters of San Benito River, S.W. of New Idria, San Benito County, California
Data: Seven, blue to blue-white benitoite xls (barium titanium silicate) associated with several large brilliant black xls of neptunite (a soda, iron, manganese titanosilicate) on and in white natrolite with greenish serpentine matrix. Benitoite xls up to ¾", neptunite up to ½"x¼". Dissolved by acid to expose xls. Size: 5"x3½"x2½". Found by W. B. Sanborn in float boulder near locality, Nov. 1958.

Name: Native Copper   No. 1212
Locality: Houghton, Keweenaw Peninsula, Michigan
Data: A typical example of a "copper skull" formed by arborescent native copper around glacial pebble. Thin walled, 1/16" thick, 4"x3"x2¾". Open on one end giving geodal appearance. Purchased, 1959.

Name: Malachite   No. 75
Locality: Apex Mine, Saint George, Washington County, Utah
Data: A basic carbonate of copper. Stalactite of dark green concentric banded malachite. Attractive and undamaged. Size: 2"x¾"x½". Traded, 1948.

Name: Stibiconite   No. 275
Locality: Antimony, Garfield County, Utah
Data: A hydrated oxide of antimony. Dull, earthy, yellowish brown pseudomorphs after stibnite showing xl form. 1½"x1"x1". Purchased 1949.

Name: *Fluorite* 544
Locality: *Heights Mine, Weardale,*
Durham, England
Data: *Calcium fluoride. Fine group of translucent sea green xls, many penetration twins with single cubes up to 1". Old Dana locality. Size: 5"x2½"x1¾", very showy, undamaged. Gift of Mrs. Joan L. Sanborn, 1956.*

Name: *Vanadinite*
Locality: *Vulture Mine,*
Castle Dome Mining District,
Castle Dome, Yuma County, Arizona
Data: *Brilliant coating of xls up to ⅛" of barrel-shaped vanadinite on rock matrix. Xls of a medium olive color with all terminations orange; unusual color combination. Associated with minor calcte. Size: 2½"x2"x1". Gift of Mr. and Mrs. D. L. Casey, 1957.*

Name: *Azurite* No. 637
Locality: *Copper Queen Mine,*
Bisbee, Cochise County, Arizona
Data: *Unusual botryoidal specimen, 3"x3"x1½", of dark blue, sky blue, and bluish-white azurite. Some minor malachite. A basic carbonate of copper.*

If a specimen has been broken and then repaired by glue, or if the crystals were loose on the matrix and have been stabilized by adhesive, then this must be indicated on the catalog entry for the specimen. Likewise, if the specimen is a cleavage, not a natural crystal, it should be indicated. A note should also be made on a catalog entry if the specimen has any polished sides or faces.

Needless to say, specimens that have been "manufactured" by gluing fragments together, or crystals onto a matrix, are fraudulent and not worthy of collection space other than as curiosities. Strangely enough, there is a slight chance of being fooled by a fake or "paste-up" specimen. A number of fakes have found their way into American collections, some so cleverly executed that even dealers have been fooled by them.

LABELING: In addition to the very important master catalog, most collectors label their specimens for convenience of identification and display. Whether the collection is housed in drawers, specimen boxes or glass cases, labeling may be done profitably.

Labeling consists of preparing an individual collection "label" for each specimen on a small printed, typed or handwritten card on which is entered the collection number (accession number), name of the specimen and locality data. Some may include the chemical formula and the date collected or acquired. A label does not carry the detailed data appearing on the catalog entry. Usually the name of the mineral, or minerals, accompanied by basic locality information is sufficient.

Professional labels usually bear the name of the institution owning the collection, or the private collector's, printed at the top or bottom of the label. A label might read: *Geology Department, University of California;* or *John A. Jones Collection;* or *Private Collection of A. E. Smith.* Labels are placed under the specimen. if housed in drawers, or beside or in front of it for display. The size of a label is a matter of preference and is determned by the size of the specimens involved, how the label is to be displayed or located, and the collection housing. Label size should be kept to a minimum, and usually a format of about 1½"x2" is adequate.

Label format is quite simple and abbreviated with the collection number (Coll., Col. No., No. or #), name (Name or Specimen), locality (Loc. or Locality), basic and standardized. Composition or properties (Comp. or Prop.), Dana number (Dana No.) and the mentioned collector or collection name are optional. Here are some typical labels of different sizes and varying format.

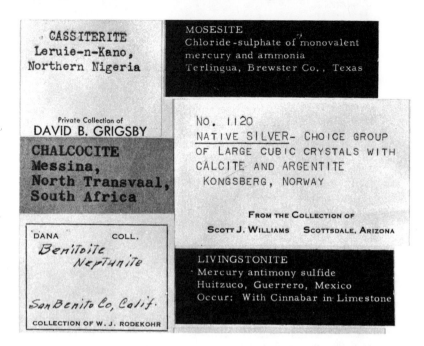

Labels may be entirely hand lettered, printed to order, or obtained from dealers in commercial pre-printed stock. Many labels from old collections are simply handwritten slips of paper. Printed labels are normally run in sheets of 10 or 12 which makes them easy to insert in a typewriter. The labels are cut out with scissors after the data has been typed on them. Commercial labels are available in different sizes and slightly differing data format. Very handsome labels may also be made by typing them on white paper, and then having the sheet of labels reproduced on a photostat negative. This yields white characters on a black background and the labels are then cut out. These are particularly effective for showcase displays.

Labels should be printed or written on a very stiff paper or thin card stock, preferably 100% rag stock for purposes of longevity. Ordinary paper stock will yellow with age. Typewritten labels produce a neat and easily read copy. If you include the chemical formula on a typed label this necessitates a typewriter with chemical symbols on the extra keys to make the job look best, or a regular typewriter equipped with a platen that may be half-spaced. A neatly hand-lettered pen-and-ink label is quite acceptable, and some collectors use special lettering devices, as do museums, which give excellent results.

The majority of labels are printed in black but there are some variations. The color of the paper stock used may be of a pastel shade. such as buff, green, or salmon, or the printing on the stock (the border and printed words) may be in a color.

Never throw away an old label accompanying a specimen. Staple it to your card catalog entry or behind your own label placed under the specimen. Old labels are very interesting and many collectors have a series of labels that have followed a particular specimen through various ownerships for one hundred or more years. When purchasing a specimen, always ask for a label with it.

In labeling or cataloging specimens on which two or more crystallized minerals are present, the most dominant, important, or spectacular mineral should be listed first. A label reading, "Tourmaline, with Quartz and Albite, Himalaya Mine, Mesa Grande, San Diego County, California" lists the minerals on a particular specimen in order of their relative importance or dominance for the collection item. If the same specimen label were to read "Albite, Quartz, and Tourmaline" it would indicate the albite crystals were dominant in this specimen, even though the tourmaline is a more prized mineral.

NUMBERING SPECIMENS: Each specimen in a collection should be clearly marked with some type of numbering system that identifies the specimen in the collector's catalog.

The number should be located in an inconspicuous spot, such as the underside or backside of a specimen. Poorly located numbers, which show when on display, detract from the appearance of the specimen. The number is essential for reference and identification purposes, but a collection should display specimens, not numbers.

There are several methods of numbering a mineral collection. The most practical and easiest is a common numerical sequence. As specimens are acquired they are given an "accession number" in normal counting sequence, beginning with number 1.

Each specimen should have its own number, but some collectors vary this by giving all specmens of the same mineral from a certain locality the identical basic number and then add an "a", "b", etc., for each specimen of this type listed. For example, if specimen No. 13 is aragonite from near Morro Bay in San Luis Obispo County, California, and the collector has four specimens of this mineral from this exact locality, he may number them 123-a, 123-b, 123-c and 123-d. The disadvantage involved is that a master card catalog or ledger never really indicates exactly how many specimens are contained in a collection since numbers may be modified and stand for two, four, or perhaps thirty specimens. For this reason most collectors give an individual accession number to each specimen. Also, printed numbers in this system are not available unless ordered specially.

There are three satisfactory ways to number specimens; (1) pen and ink; (2) printed numbers; (3) typed numbers. Whatever method is used, care should be taken always to underline any numbers that can be read backward or upside-down. For example, number 99 can be mistaken for 66. Also, adhesive tape, cellulose tape, masking tape and gummed labels are unsatisfactory — they dry and fall off.

The most permanent way to number mineral specimens is also the most time consuming, and this is with india ink on a black or white painted dot. On the reverse or backside of a specimen a round or oval dot is painted with quick-drying enamel. It is sometimes difficult to find a place where a single application of paint will yield a good numbering surface, so two or three coats are sometimes necessary to build up such a surface. The specimen is then numbered neatly by hand, and when dry, the number and the dot are coated with clear nail polish.

The majority of collectors use printed or typed numbers. Pages of printed numbers are carried in a variety of sizes by most stationery stores and many mineral dealers. Printed numbers may be obtained that are smaller than either pica or elite type of the typewriter. Printed numbers are inexpensive, costing less than one dollar for numbers up to 4000.

Numbers may also be typed. Use a fresh ribbon and fairly stiff paper or lightweight cardboard. Most collectors use white paper or cardboard for their numbers, but pastel blue, green, and yellow have also been used. The numbers, either printed or typed, are cut out carefully with sharp scissors and affixed to the specimen with a clear quick drying household acetate cement. This cement comes in a tube which makes application easy. Place a small amount of the cement on the specimen at a point where the number is to be placed, let it dry for a few seconds and press the number into the cement, then coat both number and the immediate area with cement, sealing it firmly in place. The point of a knife blade or a dentist tool are valuable aids in placing numbers on specimens.

Chapter 8

Value Of Mineral Specimens

What determines the value of a crystallized mineral specimen? Some of the attributes are obvious and others recognizable only by experienced collectors and experts. The following are important in determining specimen value:
1. The abundance of the mineral and its crystals.
2. The classification of the mineral as a gem stone.
3. The over-all quality of the crystallized specimen itself.

Discussing these will illustrate how relative and interdependent they are.

ABUNDANCE AND RARENESS: Many hundreds of the over 4,000 known minerals are very "common," such as quartz, gypsum, mica, feldspar and barite. Others are "uncommon"; they are seldom found or are very limited in locality occurrence. A third reference is to minerals which are "scarce," which usually means that at present there is an insufficient supply and implies that previously they were in greater abundance. Many minerals are classified as "rare." Rare usually means that the mineral is very infrequently found, and that it may be restricted to one or two localities, or possibly limited to a few known specimens. There are other applications of the term "rare" that are also important to the mineral collector. There are some common minerals that can occur in rare crystal forms and are, therefore, sought by collectors. An example would be stibnite. Fine crystallized specimens have been found at many localities; however, *twinned* stibnite crystals are rare.

Another application of rarity considers the infrequency of perfection in certain mineral crystals. The mineral realgar is fairly common, but perfect and choice crystals of realgar are rarely found. When found, they are usually quite small, and a one or two-inch realgar crystal would be extraordinary. The actual size of a particular individual crystal sometimes makes for rarity. Joaquinite is always found as tiny, yellow brown, orthorhombic crystals. A joaquinite crystal one quarter inch in diameter would be a giant for this mineral, and quite rare. A rare specimen does not necessarily have to be one of beauty or showiness. Many minerals, although occurring well-crystallized, are basically unattractive. For example, the crystals of xenotime from the old silica mine near Nuevo, Riverside County, California, are uncommon specimens occurring in excellent crystals — but they are an unattractive dull tan color.

Bordering on rarity are the singular and unusual crystal specimens, *i.e.*, the odd and unique of the mineral kingdom. Quartz crystals are one of the world's most prolific crystallizations, but coupled with this abundance are some unusual finds in crystals which are highly modified, oddly terminated, "twisted," enclose spectacular inclusions, or combined in a peculiar manner with some other minerals. Similarly, the various pseudomorphic forms encountered may be prized not only for their beauty but for the mineralogical story they tell.

There are a few minerals which are comparatively abundant, yet their crystals seldom occur and are rare. A good example is the mineral turquoise. Turquoise is found in limited quantities at many localities in the world, but despite its distribution, crystals of tourquoise are extremely rare.

Another example is the common element, lead. Lead is one of the most extensively mined substances in the world, and such workings are usually referred to as "lead mines." Actually, they are mines where *lead ore,* such as galena, is obtained, and from which lead is extracted. Pure native lead is rare even in massive form, and crystals of native lead are highly prized by collectors and museums alike.

Most crystallized specimens are found in mines, and this automatically creates a limited supply of specimens. With the exception of gem stone mines, few mines are operated specifically for obtaining crystals. Therefore, the crystal production of a mine is purely incidental. As a mine is worked through crystal-bearing zones, the number of specimens encountered and collected may be either limited or numerous. In several instances, outstanding specimens have been located in one pocket, vug or vein, and that was all to be found!

Specimens of choice quality found under such limited conditions can make them more valuable. A typical example is the fine pale blue topaz from the gem-bearing pegmatites near Jacumba, California. Comparatively few were found and today their value is great. In other words, certain limited locality finds can produce specimens of higher-than-average value and interest due to this limitation.

Another type of specimen which may be considered rare are those which show an unusual combination or association of minerals. Such specimens are usually more meaningful to experienced collectors who are sensitive to the possible and infrequent combination of the minerals in question. From the mineralogical standpoint, there are many combinations that seldom occur; and when such specimens are found, they are frequently of great scienitfic interest.

There is always one aspect of mineral rarity that is particularly intriguing. Rareness is always subject to change, depending on new "finds." At present, native lead crystals are rare; however, if some new mine is opened where they occur in great profusion, the degree of rarity would soon decline as such specimens began to appear in collections around the world.

In summary, the rarity of a specimen is directly proportional to its abundance in the world, its crystal properties, limited finds, and unique forms.

CLASSIFICATION AS A GEM STONE: A second value consideration is whether a specimen is a gem stone mineral or not. Gem stone crystal specimens can be worth hundreds, or even thousands of dollars each. A gem stone (either precious or semi-precious) is a material used for jewelry or ornamental purposes and is therefore in commercial demand. Gem stones have a consumer market value which transcends that of an uncut specimen. As might be assumed, the majority of mined gem stones are sold to professional lapidaries and cutters, and then to the jeweler and his customer. For jewelry, gem stones must be as flawless as possible, and even in the finest gem stone mines, flawless material is rare, a factor which keeps gem stone prices high. For every carat of facet-quality gem tourmaline mined in San Diego County, you may be sure that many pounds of unsuitable tourmaline were also taken out. Specimens that are not of cutting quality often find their way into collectors' cabinets. The numerous gem mines of San Diego County have furnished many superb collection specimens of tourmaline, beryl, kunzite, garnet and topaz.

OVER-ALL QUALITY: The third basic consideration in specimen value is the over-all quality of the specimen itself. The general attractiveness and observable perfection of a specimen is the most important aspect to the average collector, yet this is a difficult consideration of specimen value to describe. Actually, observing specimens in fine collections is the best way to gain an appreciation of what is regarded as an average, good, outstanding or superb specimen. Every specimen is different. No two specimens, even of the same mineral and from the same locality, are exactly alike in size, shape and form. There are, however, a few specific points which guide museum curators and experienced collectors in their evaluation of quality and value.

The point of greatest single importance is the "general condition" of the specimen. The specimen should be well crystallized, and the crystals should be sharp, not broken, chipped or otherwise damaged. The specimen should be free from distracting stains and extraneous matrix.

The attractiveness of a specimen (if the mineral is of the type that is showy and occurs well-crystallized), is essentially an aesthetic consideration. It is highly respected and greatly affects the value of specimens. Mineral specimens can be of extraordinary beauty. Even common minerals, such as calcite, barite and gypsum, frequently occur in specimens of spectacular beauty and consequently are prized by collectors for this attribute. Specimens often show two, three, or more different crystallized minerals; and such combinations can make a more attractive and desirable specimen.

Frequently, well-crystallized specimens of high quality will lack pleasing format or organization as a collection or display specimen, and therefore are not as attractive as others. Perhaps the specimen is of ungainly shape, or the crystals are so located that the specimen is difficult to display. A group of equal crystal quality, but of more attractive proportions would be more desirable and useful as a collection specimen and therefore more valuable.

MARKET: Is there a demand or market for crystal specimens? The answer is yes. Private collectors, schools and museums purchase quantities of specimens. The hobby of mineral collecting has greatly increased in popularity in the last few years. There are dozens of established mineral dealers and natural-science supply houses in various parts of the country.

TRADING: It is virtually impossible for any one collector to collect all the various specimens he desires, so he frequently buys or trades specimens. Many fine collections have been established by trading. A collector may build up a reserve of good duplicate specimens from his immediate area. With these he can trade with other collectors. He can also sell directly to a dealer, or perhaps take out an amount in trade at the dealer's.

PRICE: How much specimens are worth in dollars is a flexible matter. One can pay from fifty cents to over one hundred dollars for a good specimen. It all depends on how common or rare the mineral is, the quality and size of the crystals, and attractiveness of the individual piece itself. It is a matter based on a relative consideration of all the factors previously mentioned in this chapter, *plus experience*.

Experienced collectors generally know the current market value of a specimen, or its approximate fair price. However, price is subject to change. In a very short period of time specimens that once were relatively common and reasonably priced may become scarce. There are many cases where a deposit yielding fine specimens has been exhausted, or the mine closed down, thereby terminating the supply.

In the 1920's, during the operation of mines in the gem stone bearing pegmatites of San Diego County, handsome cabinet specimens of tourmaline, quartz, and albite could be obtained for a few dollars. Ten to twenty dollars would buy excellent and outstanding collection specimens. Most of these mines are no longer in operation. Today such specimens are worth many times the original value. Although fine quality specimens always command a fair price, they are worth even more if the locality is closed down or worked out. This obviously affects prices.

Much the same situation applies to the famed Tsumeb locality in South West Africa. In the early 1900's spectacular specimens of huge crystals of azurite, malachite, smithsonite, cerussite and related minerals came from the mines of this locality. Then the workings passed through the crystal-producing zones, and consequently the specimens began to assume a new value aspect since the supply had been terminated. However, in recent years the mines at Tsumeb have once again been the source of such specimens. Although they are the same minerals, the new finds have a much different appearance from the earlier specimens. New material from Tsumeb is available at present, but it, too, may be terminated by the direction and depth of mining operations. A fine "old Tsumeb specimen" is still, and probably always will be, a choice item.

Collectors are often prone to think that, because a locality produces quantities of material at the moment, this material is of little value. This is particularly true of some collectors' attitude toward the great variety of truly excellent crystallized minerals from the well known Tri-State District around Joplin, Missouri. These mines will not operate forever, and in the years to come, their fine specimens will increase in value despite the fact that thousands of them are available today.

BUYING SPECIMENS: Generally speaking, the price of mineral specimens is proportional to the supply and demand. Considering the average hobbyist, and even the advanced collector, the following approximate prices may be cited. They are, of course, completely relative to the size of the specimen in question, and are not applicable to rare or unique specimens. With common or fairly common minerals, of which hundreds occur well-crystallized and in ample supply, good specimens may be obtained for fifty cents to two or three dollars; very good specimens may range up to five or ten dollars, and specimens priced from ten to twenty dollars and up should be outstanding in all respects. Prices vary greatly from dealer to dealer, according to his source of supply, his overhead and desired profit. It is a fallacy to believe specimens can always be purchased cheaper at the mine, quarry, or in the locality area. Bargains can be obtained this way, but frequently prices charged are outrageous. For example, gold nuggets at $75 an ounce with a gold value of $35 an ounce do not make too much sense. There are, however, a few wealthy collectors who will pay almost any price asked for a desired specimen. Consequently a base price is established which is beyond the average collector.

Spending money on fine quality specimens is actually an investment. Very little, if any, loss is taken by the buyer, provided the specimen is well chosen and the price paid is within reason. When one purchases a fine mineral specimen an investment is also made in individuality; although there may be hundreds of similar specimens, you can rest assured there is not another exactly like yours.

Specimens in sizes 4"x5" and smaller are in heaviest demand. Large specimens are not so intensively sought by hobbyists, and their market is limited to a few collectors, colleges, universities and museums. Also, specimens of large size and of high quality can be quite expensive.

Specimens which meet the various aforementioned criteria are not especially difficult to obtain or find. A collection is just as valuable as the quality of its specimens and its meaning to the collector. It would be hard to place a value on the pride that a collector rightly feels if he has personally found the majority of specimens in his cabinets. There are also the fine recreational experiences and associations gained while hunting specimens along the seashore, in the deserts and the mountains.

Here are a few general pointers to keep in mind when purchasing specimens:

1. Stay within your budget limitation for specimens.

2. Be selective within your budget. Get the best specimen you can for what you care to pay. For example, if a dealer displays a tray of wulfenite crystals from the Ojuela Mine, Mapimi, Durango, Mexico, all priced at $1.00 per specimen, you may be sure that some are superior to others at the same price.

3. Buy specimens of attractive format which will display well for your purposes.

4. Be sure purchased specimens are accompanied by a label.

5. If you deal by mail, be sure you have an understanding with the dealer that you may return any specimens with which you are not satisfied. Most dealers operate on this basis.

6. Do not purchase damaged, bruised, chipped, ungainly or generally "beaten up" specimens, unless you see a bargain involved or the possibility of trimming the specimen into a more acceptable size for your purposes.

7. Stay with a particular mineral size, or sizes, in your buying. The over-all consistency of size within a collection is desirable.

8. Visit dealers' stores and displays at mineralogical shows and note specimen prices for purposes of comparison. This is the best way to learn the "market" as well as gain some idea of what your own specimens may be worth.

9. Keep the quality of your specimens as high as possible. Purchasing one or two fine specimens during a visit to a dealer is probably better than investing the same amount of money in eight or ten average specimens.

Generally speaking, mineral specimens of good quality are a sound investment. Unless you pay an exorbitant price for a specimen, it will in all probabilty increase in value over the years.

OTHER GEMBOOKS FOR GEM AND MINERAL HOBBYISTS

CABOCHON CUTTING. A Gem Cutter's Handbook by Jack R. Cox. Profusely illustrated, step-by-step instructions show you how to cut cabochon gemstones and set them in a variety of jewelry mountings. **$2.00**

ADVANCED CABOCHON CUTTING. A Gem Cutter's Handbook by Jack R. Cox. Shows you how to cut special shapes, star gems and cat's-eyes. Special chapters on opal, jade and assembled stones. Well illustrated. **$2.00**

SPECIALIZED GEM CUTTING. A Gem Cutter's Handbook by Jack R. Cox. Learn how to finish flat surfaces, drill, tumble, make spheres, bookends, etc. Special section shows how to work with diamond abrasives. **$2.00**

THE ART OF GEM CUTTING by H.C. Dake. Over 70,000 in seven editions have started more beginners than any other book. Covers basic gem cutting, faceting, spheres, flats, etc. **$2.00**

FACET CUTTERS HANDBOOK by Edward J. Soukup, G.G., F.G.A. A complete step-by-step guide that tells and shows how to cut faceted stones of many types. Detailed drawings and instructions by an expert teacher. **$2.00**

HANDBOOK OF GEMSTONE CARVING by Ed and Leola Wertz. A complete guide for amateurs with step-by-step instructions for carving flat work, carvings in the round and portraits in stone. Profusely illustrated. **$2.00**

HANDBOOK OF LOST WAX OR INVESTMENT CASTING by James E. Sopcak. Shows you how to make and use equipment to produce patterns, molds and castings for jewelry and small metal parts. Completely illustrated. **$2.00**

CHANNEL JEWELRY — A NEW APPROACH by Stanley Tims. An up-to-date technique which simplifies making beautiful Zuni Indian style jewelry. Illustrated loose leaf sheets fit in a standard ring binder. Postpaid. **50¢**

HANDBOOK OF JADE by Gerald I. Hemrich. Written specifically for gem hobbyists. Covers jadeite, nephrite, chloromelanite, pseudojades; where to find, how to identify, cutting characteristics and techniques. **$2.00**

HANDBOOK OF CRYSTAL AND MINERAL COLLECTING by William B. Sanborn. This guide tells you how and where to look for minerals and crystals as well as how to clean, prepare, catalog and display your collection. **$2.00**

THE COMPLETE GUIDE TO MICROMOUNTS by Milton L. Speckels. A complete manual and how-to-do guide to the fascinating world of microminerals. How to collect, work with, mount and display micromounts. **$2.00**

DESERT GEM TRAILS by Mary Frances Strong. The most detailed guide to the gem, mineral, fossil and scenic treasures of the Mojave and Colorado Deserts of California. Each area mapped, with mileages, etc. **$2.00**

MIDWEST GEM TRAILS by June Culp Zeitner. A detailed field guide to the gems, minerals and fossils of 12 Midwestern states. Illustrated with pictures and maps. Hundreds of collecting localities listed. **$2.00**

EASTERN GEM TRAILS by Floyd and Helga Oles. A charming guide to the gem and mineral localities of the middle Atlantic states with emphasis on productive and accessible areas. Good reading, good guiding. **$2.00**

FIELD GUIDE TO THE GEMS AND MINERALS OF MEXICO by Paul W. Johnson. The only detailed guide to 100's of gem and mineral localities in Mexico. Spanish-English glossary. Illustrated. Maps. **$2.00**

FAMILY FUN WITH ROCKS. A beginner's book with eleven great step-by-step rock hobby projects. Excellent for all ages. Gives a basic introduction to other phases of the gem and mineral hobby. Well illustrated. **$1.00**

JEWELRY MAKING FOR BEGINNERS — THE SCROLL WIRE METHOD by Edward J. Soukup, G.G., F.G.A. Presents an easy-to-learn technique which enables you to produce professional quality jewelry without breaking the bank. **$2.00**

JEWELRY MAKER'S HANDBOOK — By Iva L. Geisinger. One of the most complete books on jewelry making available. Covers tools, wire work, piercing, soldering, surface texturing, finishing, designing and lots more. **$2.00**

GEMS & MINERALS — The leading Gem and Mineral Hobby Magazine — The one that tells you with pictures and words how to get the most from the gem and mineral hobby. **Only $6.50 per year; $12.00 for two years.**

Please add 25¢ postage on book orders of $1.00 or more. California residents add sales tax.
FROM YOUR DEALER or GEMBOOKS, Mentone, California 92359